MOBILE MICROROBOTICS

Intelligent Robotics and Autonomous Agents

Edited by Ronald C. Arkin

A complete list of the books in the Intelligent Robotics and Autonomous Agents series appears at the back of this book.

MOBILE MICROROBOTICS

Metin Sitti

The MIT Press
Cambridge, Massachusetts
London, England

This book was set in Times Roman by the author. Printed and bound in the United States of America.

Library of Congress Cataloging-in-Publication Data:

Names: Sitti, Metin, author.
Title: Mobile microrobotics / Metin Sitti.
Description: Cambridge, MA : MIT Press, [2017] | Series: Intelligent robotics and autonomous agents | Includes bibliographical references and index.
Identifiers: LCCN 2016047358 | ISBN 9780262036436 (hardcover : alk. paper)
Subjects: LCSH: Microrobots. | Mobile robots.
Classification: LCC TJ211.36 .S57 2017 | DDC 629.8/932 – dc23
LC record available at https://lccn.loc.gov/2016047358

10 9 8 7 6 5 4 3 2 1

To the beautiful memories of my beloved sister, brain sergeant, İlkay Sitti
whom we lost so young and so unexpectedly

Contents

List of Figures

1.1 Diagram showing the benefits, challenges, and potential applications of mobile microrobots. 4

1.2 A conceptual sketch of an example future mobile microrobot with spatio-selective surface functionalization for potential medical applications. Each functional component could be assembled on a main body. The main body further could serve as a large depot for therapeutics to launch controlled release at the site of action. A closed-loop autonomous locomotion (e.g., a bio-hybrid design) could couple environmental signals to motility. Targeting units could enable reaching and localization at the intended body site. Medical imaging, e.g., magnetic resonance imaging (MRI), contrast agents loaded on the microrobot could enable visualization as well as remote steering on demand. Metallic nanorods could enable remote plasmonic or RF heating to decompose a tumor tissue by hypothermia. 5

1.3 Approximate timeline showing the emerging new microrobot systems with their given overall size scale as significant milestones. (a) Implantable tiny permanent magnet steered by external electromagnetic coils [1]. (b) Screw-type surgical millirobot [2]. (c) Bacteria-driven bio-hybrid microrobots [3]. (d) Self-electrophoretic catalytic microswimmer [4]. (e) Bio-hybrid magnetic undulating microswimmer [5]. (f) Glucose-fueled catalytic microswimmer [6]. (g) Magnetically controlled bacteria [7]. (h) MEMS electrostatic microrobot [8]. (i) Thermal laser-driven microrobot [9]. (j) Magnetic bead driven by an MRI device in pig artery [10]. (k) Magnetic microswimmer with rigid helical flagellum inspired by bacterial flagella [11, 12]. (l) Crawling magnetic microrobot [13]. (m) Microtubular catalytic jet microrobot [14]. (n) Bacteria swarms as microrobotic manipulation systems [15]. (o) 3D magnetic microrobot control [16]. (p) Self-thermophoretic microswimmer [17]. (q) Bubble microrobot [18]. (r) Light-sail microrobot [19]. (s) Self-acoustophoretically propelled microrobot [20]. (t) Sperm-driven bio-hybrid microrobot [21]. (u) Magnetic, chemotactic, and pH-tactic control of bacteria-driven microswimmers [22–24]. (v) Magnetic soft undulating swimmer [25]. (x) Untethered pick-and-place microgripper [26]. (y) Cell-laden microgel assembling microrobot [27]. (z) Catalytic micromotors driven by enzymatic reactions [28]. (aa) In vivo navigation of

microswimmers [29, 30]. (bb) 6-degrees-of-freedom (6-DOF) actuation of magnetic microrobots [31]. 6

1.4 Some existing remote (off-board) approaches to mobile microrobot actuation and control in 2D. (a) Magnetically driven crawling robots include the Mag-μBot [13], the Mag-Mite magnetic crawling microrobot [32], the magnetic microtransporter [33], the rolling magnetic microrobot [34], the diamagnetically levitating milliscale robot [35], the self-assembled surface swimmer [36], and the magnetic thin-film microrobot [37]. (b) Thermally driven microrobots include the laser-activated crawling microrobot [9], the micro-light sailboat [19], and the optically controlled bubble microrobot [18]. (c) Electrically driven microrobots include the electrostatic scratch-drive microrobot [38] and the electrostatic microbiorobot [39]. Other microrobots which operate in 2D include the piezoelectric-magnetic microrobot MagPieR [40] and the electrowetting droplet microrobot [41]. 8

1.5 (a) Chemically propelled designs include microtubular jet microrobots [14], catalytic micro/nanomotors [42], and electro-osmotic microswimmers [43]. (b) Swimming microrobots include the colloidal magnetic swimmer [5], the magnetic thin-film helical swimmer [44], the microscale magnetic helix fabricated by glancing angle deposition [12], the helical microrobot with cargo carrying cage fabricated by direct laser writing [45], and the helical microrobot with magnetic head fabricated as thin-film and rolled using residual stress [46]. (c) Microrobots pulled in 3D using magnetic field gradients include the nickel microrobot capable of 5-DOF motion in 3D using the OctoMag system [16] and the MRI-powered and imaged magnetic bead [47]. (d) Cell-actuated biohybrid approaches include the artificially magnetotactic bacteria [48], the cardiomyocyte-driven microswimmers [49], the chemotactic steering of bacteria-propelled microbeads [24], the sperm-driven and magnetically steered microrobots [21], and the magnetotactic bacteria swarm manipulating microscale bricks [15]. 9

2.1 Left image: Transmission electron microscope image of a multi-flagellated *E. coli* bacterium with around 0.5-μm diameter and 2-μm length. Copyright © Dennis Kunkel Microscopy, Inc. Right image: Photograph of an example scaled-up multiple bacterial helical flagella setup for measuring the thrust force produced by multiple flagella as a function

of flagella geometry and distance. A planetary gear system was used to guarantee individual rotation of each flagella at identical rates. Reprinted from [50] with the permission of AIP Publishing. 23

3.1 Schematics of surface tension forces acting at the liquid-solid-vapor interface, which result in a thermodynamic equilibrium contact angle θ_c on a solid surface in a given vapor such as air. 29

3.2 Schematics of liquid droplet wetting on structured surfaces: (a) Cassie regime where the liquid droplet does not fully wet the micro/nanostructure and there is air trapped under droplet, and (b) Wenzel regime where the droplet fully wets the micro/nanostructure. 31

3.3 Short- or long-range, contact or non-contact, and attractive or repulsive surface forces, such as van der Waals, electrostatic, and capillary forces, create a sticky world for a microrobot interacting with other surfaces or objects in a given medium. 32

3.4 Schematics of parameters during a capillary force between a spherical microrobot asperity and a flat, smooth substrate due to the liquid bridge between two surfaces in air. 37

3.5 Schematics of electrowetting of a conductive liquid droplet on a flat, smooth dielectric film. 38

3.6 Comparison of weight and adhesive forces for a gold sphere in close 0.2 nm contact with a gold surface. The surfaces are assumed atomically smooth gold, with a Hamaker constant of 400 zJ for gold. The medium is assumed as air, with a relative permittivity of 1. The sphere was assumed to be made from gold ($\rho = 19,300$ kg/m^3) for the weight calculation, and to have a voltage of 100 V for the electrostatic calculation in an air environment. For the capillary force, $\gamma = 0.0728$ N/m is chosen for water with $\theta_c = 85°$ for water contact angle on gold. 41

3.7 Pressure (stress) profiles for (a) Hertz, (b) JKR, (c) DMT, and (d) MD elastic contact mechanics models, where compressive (positive sign) external normal load stress and tensile (negative sign) adhesive stress create small deformation on a spherical object in contact with a flat plane. 48

3.8 Contact geometry of a sphere deforming on a flat plane by (a) Hertz, JKR, and DMT models; and (b) MD elastic contact mechanics model for an applied external load of L and attractive surface forces increasing the

contact deformation differently depends on the contact model. The MD theory models the periphery of the sphere-plane interface as a crack. 49

3.9 Side-view sketch of a sphere rolling on a plane due to an applied rolling moment M_r. Top view of the approximated shape of the rolling contact for a sphere on a plane. 56

3.10 Schematic of an example custom force measurement setup to characterize adhesion and friction at the micron scale for mN scale interaction forces: A - load cell, B - (spherical, circular flat punch, etc.) indenter, C - flat surface, D - microscope objective, E - two-axis manual linear stage, F - two axis manual goniometer, G - motorized linear stage, H - light source. 62

3.11 A possible force-distance curve between a smooth glass hemisphere loaded to and unloaded from a smooth, flat substrate. Two surfaces start approaching to contact from point A. Long-range attractive forces could attract the hemisphere with a maximum attractive force at point B. Two surfaces contact each other at point C when the distance reaches the interatomic distance a_0. The interface would deform elastically during compressive loading until reaching a predetermined load L at point D. An attractive (negative) maximum tensile force (i.e., pull-off force P_{po}) occurs at point E. Two surfaces separate from each other at point F. 63

3.12 (a) Single polymer fiber with a spherical tip adheres to a flat glass substrate (left image), and (b) two polymer fiber spherical tips are in contact (right image). 66

4.1 Scanning electron microscope (SEM) image of an example (a) magnetic microrobot fabricated by laser micromachining from a bulk NdFeB sheet to achieve strong magnetization properties, and (b) polymeric microrobot (coated with a magnetic Cobalt nanofilm) 3D-printed by a two-photon lithography system to trap a microbubble on the hole on the upper side of the cubic robot body for picking and placing small parts using capillary forces inside fluids. Scale bars are 200 μm. 70

4.2 An example 3D microrobot design fabricated by two-photon lithography. (a) Each component can be selectively addressed by the two-photon laser pulse, thereby allowing for selective functionalization via light chemistry. The catalytic engine wall is selectively patterned with platinum nanoparticle groups that produce gas microbubbles inside the engine.

The microbubbles leave the robot through the nozzle, inducing a thrust that propels the microrobot forward. (b) Section view of the catalytic microrobot shows its components in a 3D CAD drawing. (c) SEM micrograph of an example microrobot fabricated from a biocompatible polymer, polyethylene glycol diacrylate. 74

4.3 Replica molding process based on photolithography used to fabricate a large number of magnetic microrobots. From top left to bottom right, process steps are: deposit an SU-8 photoresist on a silicon wafer; pattern the SU-8 layer using UV lithography; replicate the SU-8 pattern's negative using a silicone rubber mold; mold the rubber mold with a liquid polymer mixed with magnetic microparticles and use a punch to remove the excess polymer mix from the mold; after curing the polymer, demold the microrobots. This process allows for the creation of arbitrary 2D-shaped polymer composite microrobots from micron to mm scale. 75

4.4 Fabrication and magnetization process for two magnetic microrobots with an integrated compliant flexural gripper [26]. Copyright © 2014 by John Wiley Sons, Inc. Reprinted by permission of John Wiley & Sons, Inc. (a) A magnetic slurry consisting of magnetic microparticles and polymer binding matrix is poured into the negative mold. (b) Microgripper shapes are pulled from the mold using tweezers. (c) Torque-based designs are spread open prior to magnetization, to allow each gripper tip to be magnetized in an opposite direction. The bend direction shown results in a normally closed gripper. Force-based microgrippers are molded from two magnetic materials, in two separate molding batches. The pieces are fixed together using UV-curable epoxy with a rubber mold as a fixture to hold the parts precisely. These force-based gripper tips are magnetized in one common direction. (d) After relaxation, the grippers are shown in their final magnetic configurations. (e) Fabricated designs are shown in the relaxed state after magnetization and assembly. 77

4.5 Fabrication, magnetization, and actuation processes of a swimming-sheet magnetic soft microrobot. Reprinted from [25] with the permission of AIP Publishing. (a) A flat sheet fabricated from permanent magnetic microparticles and Ecoflex silicone rubber is (b) bent into a circle and subject to a 1.0 T uniform magnetic field. (c) When the field is removed and the elastic robot body is straightened, it is left with a magnetization

that varies along its length, which (d) causes it to be deformed when subject to a weak external field. Rotating the external field continuously in time causes the sheet deformations to travel down its length, providing a propulsive force in fluid. 78

4.6 Scanning electron microscope images of self-folded untethered microgrippers from 2D patterns using directed self-assembly. Reprinted from [51]. All rights reserved. 81

5.1 Conceptual drawing of a microrobot capable of sensing and interacting with entities within its environment. In this example, the microrobot could be capable of detecting chemicals through the use of chemical sensors and could be able to transmit these data wirelessly. A microrobot may also have a manipulator to actively control objects in its environment. 86

5.2 Schematic of a possible sensing scheme used for detecting changes in natural frequency for a magnetoelastic sensor (MES). An electromagnetic pulse can be sent using the drive coil, causing the magnetoelastic sensing element to oscillate. This oscillation would create a decaying electromagnetic signal, which could be picked up by the sensing coils. The electromagnetic response could be analyzed to determine the amplitude and resonant frequency of response. 94

6.1 Basic directions in a 3D anisotropic polycrystalline piezo ceramics. 99

6.2 (a) Axial (left image) and (b) transversal (right image) type piezo actuation. 100

6.3 Basic cantilevered rectangular-shaped piezoelectric unimorph actuator design for small-scale actuation. 102

6.4 Fixing the PZT-5H piezo layer dimensions and applied voltage, the elastic layer's thickness h_s can be tuned to maximize the displacement or mechanical energy output of the unimorph actuator. 104

6.5 Linear dynamic model (bottom image) of a flapping wing design (upper image) with a piezo actuator, lossless transmission, and wing. 105

6.6 Serial and parallel type bimorph piezo actuators, where the two active piezo layers are bonded to each other with the given poling directions (the arrow in each piezo layer) and electrical connections. The

upper layer expands and the lower layer contracts for enhanced bending motion. 108

6.7 Dielectric elastomer actuator contracts vertically and expands laterally when high voltage is applied at its compliant electrodes. 115

6.8 Side-view sketches of two thermal microactuator designs: (a) One-layer metal cantilever beam bends when heated because each beam arm heats up and expands differently due to their significantly different geometry and electrical resistance. (b) Bi-layer cantilever beam consists of two layers with significantly different coefficients of thermal expansion. 117

6.9 Top-view (above) and side-view (bottom) drawings of an electrostatic comb-drive microactuator design with *N* fingers. 118

6.10 Basic operation modes of MRF and ERF actuator designs: (a) flow, (b) shear, and (c) compression modes. 119

7.1 Self-electrophoresis, where E is the electric field and H^+ and e^- show the direction of flow of ions and the electric field, respectively. 124

7.2 Self-diffusiophoresis, ∇C, the concentration gradient, inducing the pressure gradient ∇P to propel the microrobot body. 127

7.3 Main components of a bio-hybrid microrobot [52]. Copyright © 2014 by John Wiley Sons, Inc. Reprinted by permission of John Wiley & Sons, Inc. These microrobotic systems integrate biological cells with artificial substrates to provide actuation and sensing functionalities. 134

7.4 Stochastic swimming motion behavior of *S. marcescens* and a *S. marcescens*-propelled microswimmer [52]. Copyright © 2014 by John Wiley Sons, Inc. Reprinted by permission of John Wiley & Sons, Inc. (a) A free-swimming bacterial cell alternates between run states, where it travels in a straight line (red arrows), and tumble states, where it tumbles and reorients in 3D space (blue dots). (b) Experimentally measured 3D swimming trajectory of a free-swimming bacterial cell (*S. marcescens*). (c) Propulsive forces and torques generated by a single bacterium attached to a microbead. (d) Representative 3D helical trajectory of a *S. marcescens*-propelled microbead, which was obtained experimentally [53]. 140

7.5 Control strategies for a bacteria-propelled microswimmer [52]. Copyright © 2014 by John Wiley Sons, Inc. Reprinted by permission of John Wiley & Sons, Inc. Experimental time-lapse images are shown on the

right for microswimmers in (a) an isotropic environment resulting in stochastic motion; (b) a linear chemoattractant (L-aspartate) gradient, which is sensed by the bacteria, leading to a biased random walk; and (c) an applied uniform magnetic field of 10 mT resulting in directed motion. The bacteria are attached to a superparamagnetic microbead in (c) and to polystyrene microbeads in (a) and (b). The red and blue lines indicate the trajectories of the microswimmers. Different microswimmer samples are shown for each environmental condition. Scale bars: 20 μm. 142

7.6 Free-body diagram of the swimming bio-hybrid microrobot, where $\vec{F}$ and $\vec{T}$ are the instantaneous total bacteria propulsion force and torque, $\vec{f}$ and $\vec{\tau}$ are the hydrodynamic translational and rotational drag on the bead, and $\vec{v}$ and $\vec{\omega}$ are the bead translational and rotational velocity vectors, respectively. 146

7.7 Sample 2D stochastic trajectories of bacteria-propelled beads with 5 μm diameter from (a) experiments (reprinted from [53] with the permission of AIP Publishing), and (b) simulations based on the computational stochastic bead motion model in Eqs. [7.8-7.9] for single and multiple (up to 15) attached *S. marcescens* bacteria. 147

8.1 An example eight-coil system capable of applying 5-DOF magnetic force and torque in a several-cm-sized workspace with an uniform magnetic field. This system is capable of applying fields of strength 25 mT and field gradients up to 1 T/m using optional iron cores. A: Top camera. B: Side camera. C: Magnetic coils. D: Workspace. 159

8.2 An example non-uniform magnetization profile to achieve 6-DOF magnetic actuation [31]. Copyright © 2015 by SAGE Publications, Ltd. Reprinted by permission of SAGE Publications, Ltd. Here, the robot has a net magnetization $\vec{m_e}$ along its local z-axis, and the magnetization vectors (solid vectors) are always pointing away from the origin. When a spatial gradient $\partial B_y/\partial x$ is applied, the induced forces on the magnetization vectors, indicated by the dotted vectors, exert torque around the z-axis of the robot body. 162

9.1 Basic operation schematic of a conventional fuel cell. 173

10.1 Top-view picture of an example 500 μm star-shaped Mag-μBot pushing a plastic peg into a gap in a 2D planar assembly task. The arena width is 4 mm. 188

10.2 Schematic of a rectangular magnetic microrobot with applied external forces and torques. Here, the typical dimensions are several hundred microns on a side, and the microrobot is made from a mixture of NdFeB magnetic powder and a polyurethane binder. The magnetization vector is denoted by $\overrightarrow{M}$. The external forces include the magnetic force and torque F, T, the fluid damping force and torque L and D, the friction force f, the adhesion force F_{adh}, the weight mg, and the normal force N. 190

10.3 Example comparison of experimental and simulation microrobot speed values using stick-slip motion on a flat silicon surface [13]. Copyright © 2009 by SAGE Publications, Ltd. Reprinted by permission of SAGE Publications, Ltd. Average microrobot speeds are given for operation in two different operating environments: air and water. Error bars denote standard deviation in experimental results. 191

10.4 Mag-μBot force scaling from the case study. Equivalent forces are computed from torques by dividing by the microrobot size. The fluid environment is assumed as water with viscosity $\mu = 8.9 \times 10^{-4}$ Pa·s, and the microrobot density is 5,500 kg/m^3. The microrobot velocity is 1 mm/s, and its rotation rate is taken as swinging through an angle of 40° at a rate of 50 Hz, or about 70 rad/s. The magnetic field is taken as 6 mT and the field gradient as 112 mT/m, with a microrobot magnetization of 50 kA/m. To calculate surface friction, the interfacial shear strength is taken as one-third the shear strength, as $\tau = 20$ MPa, and the contact area is varying with load as given in Section 3.5. The gap size for adhesion calculation is taken as 0.2 nm. The coefficient of friction μ_f is taken as 0.41. The work of adhesion W_{132} is calculated in water for the polyurethane and silicon surfaces and is found to be negative, indicating repulsion. This material pairing was chosen specifically to yield this negative value. This results in a steep drop in the friction force when the microrobot weight overcomes this repulsive force at a microrobot size of about 7 mm. In a model with non-smooth surfaces, the friction would be positive at smaller scales. 193

10.5 Microscale swimming methods. (a) Rotation of a stiff helix inspired by bundled flagella of swimming *E. coli* bacteria. (b) Traveling wave through an elastic tail or body inspired by spermatozoids. 196

10.6 Side-view sketch of a cylindrical microrobot body with fluid contact angle θ_c staying on a fluid-air interface in equilibrium. V_b is the volume under water due to buoyancy, and V_{st} is the volume under water due to surface tension F_{cap}. 201

10.7 Numerically estimated maximum lift forces for different fluid contact angles (θ_c) of an example robot body with the cylindrical geometry shown in the inset image, showing that the robot body should be hydrophobic to have high surface tension-based lift forces [54]. Copyright © 2007 by IEEE. Reprinted by permission of IEEE. 202

10.8 Example milliscale mobile robots that walk on water inspired by water strider insects: (a) a 0.65-gr tethered water walking robot with four water-repellent supporting legs and two driving legs actuated by three unimorph piezos at resonance (left photo), and (b) a 22-gr water walking robot with 12 water-repellent circular concentric feet and two driving feet actuated by two tiny DC motors (right photo). 204

11.1 Components of feedback control of a general microrobotic system. 212

11.2 Conceptual sketch of a multi-robot control system, where a large number of magnetic microrobots could be remotely actuated and controlled or self-propelled autonomously to achieve a variety of tasks inside the human body or other operation environments [55]. Copyright © 2015 by IEEE. Reprinted by permission of IEEE. Here, such a microrobotic swarm could be addressed and controlled individually, as teams, or as an ensemble. 213

11.3 Robot velocity versus electrostatic anchoring voltage for a microrobot on a 6-μm-thick SU-8 layer. Reprinted from [56] with the permission of AIP Publishing. A critical voltage of 700 V is required to affix the microrobot. Videos of the motion were recorded and analyzed to determine velocities. A pulsing frequency of 20 Hz was used for translation. 216

11.4 Experimental velocity responses of two Mag-μBots with varying aspect ratios but similar values of effective magnetization [45,67]. The maximum field strength was held at 1.1 mT. Data points are mean values, and error bars represent standard deviations for 10 trials. 217

11.5 *H-m* hysteresis loops of microrobot magnetic materials, taken in an AGFM for applied field up to 1,110 kA/m shows distinct material coercivity values. The magnetization is normalized by the saturation magnetization M_s of each sample. 219

11.6 Schematic showing the multiple magnetic states, which can be achieved through the use of a variety of magnetic materials [57]. Copyright © 2012 by IEEE. Reprinted by permission of IEEE. (a) Three separate magnetic actuators, each made from a different magnetic material, the magnetization of which can be independently addressed by applying magnetic field pulses of various strengths. Here, H_{pulse} is a large field pulse and H_{small} is a small static field. (b) A single magnetic composite actuator can be switched between the "up," "off," or "down" states by applying pulses of different strength, where H_{large} is a large field pulse. 221

11.7 Addressable microrobot teamwork task, requiring the cooperative contribution of two mobile microrobots of different sizes working together to reach a goal [57]. Copyright © 2012 by IEEE. Reprinted by permission of IEEE. Frames show two superimposed frames, with the microrobot paths traced and midpoints outlined. (a) Both microrobots lie inside an enclosed area. The door to the goal is blocked by a plastic blockage. Only the larger microrobot can move the blockage, while only the smaller microrobot is small enough to fit through the door. (b) The larger microrobot is enabled and moved to remove the blockage while the smaller disabled microrobot remains in place. The larger microrobot is returned to its staring point and disabled. (c) The smaller microrobot is enabled and is free to move through the door to the goal. 222

12.1 (a) A teleoperated star-shaped microrobot and a 210-μm microsphere for side-pushing under liquid on a glass surface [58]. Copyright © 2012 by IEEE. Reprinted by permission of IEEE. (b) The microrobot moves past the microsphere from its side, causing the sphere to displace a small amount of D_s, primarily due to the fluid interactions. Arrow on microrobot indicates direction of its motion. 227

12.2 Simulation and experiment of a star-shaped microrobot manipulating a 210-μm microsphere from the side [58]. Copyright © 2012 by IEEE. Reprinted by permission of IEEE. Vertical division indicates whether sphere contact occurs with the microrobot's edge, determined from the simulation. The simulation fit "Sim Fit" is from the dynamic simulation,

while "Sim Fit Lin" is a linear approximation to this fit, which can be used for control using these results. 228

12.3 A side-view slice of the finite element modeling (FEM) solution for the flow around a star-shaped microrobot as it traverses through the environment [58]. Copyright © 2012 by IEEE. Reprinted by permission of IEEE. The microrobot is moving toward the left in these images, and the flow velocities correspond to y-directed flow, depicted by arrows. Half the microrobot is modeled in this analysis. 229

12.4 Team non-contact manipulation by three microrobots simultaneously spinning [59]. Copyright © 2012 by IEEE. Reprinted by permission of IEEE. The microrobot positions are trapped at discrete locations by magnetic docks embedded in the surface. Manipulated microsphere paths are tracked by colored lines. 230

12.5 Trapping and translating live cells or other microentitites by a spinning and rolling magnetic spherical microrobot [60]. Reproduced with the permission of the Royal Society of Chemistry. (a) The simulation result of a 5-μm diameter spherical microrobot spinning at 100 Hz on a flat surface in water. The plot is from a top view of the cross-section taken at the equatorial plane of the microrobot. Red concentric circles represent the streamlines. The color map shows the flow velocity distribution. (b) The schematic of the trapping of a nearby bacterium by the rotational flow induced by the rotation ω of the robot near a flat surface. Any bacterium that is far away is minimally affected by the induced flow. A bacterium that is close enough to the spinning particle is first reoriented by the flow to align its body's long axis with the local streamline (i). Then it is trapped (ii) and orbited around the particle. (c) The schematic of the mechanism for enabling the mobility of the induced rotational flow field. Instead of being perpendicular to the surface as shown in (b), the rotation axis of the particle is tilted from the z-axis. (d) The finite element simulation result of a 5-μm diameter spherical microrobot rotating at 100 Hz with a tilt angle of 75 degrees and a translational speed of $0.06\omega_r a$ in the $-y$ direction on a flat surface in water. The plot is from a top view of the cross-section taken at the equatorial plane of the robot. The arrows indicate the in-plane flow velocity at selective positions, while the color map shows the distribution of the out-of-plane flow velocity normalized by the magnitude of the in-plane flow velocity at the same position. 231

12.6 Microgel blocks with embedded live cells have been assembled using a magnetically driven microrobot on a planar surface in physiological fluids [27]. Copyright © 2014 by Nature Publishing Group. Reprinted by permission of Nature Publishing Group. Fluorescence images of NIH 3T3 cell-encapsulating hydrogels after the assembly of (a) T-shape, (b) square-shape, (c) L-shape, and (d) rod-shape constructs. Green represents live cells and red represents dead cells. (e-g) Immunocytochemistry of proliferating cells stained with Ki67 (red), DAPI (blue), and Phalloidin (green) at day 4. (e) Cells stained with DAPI and Phalloidin at 20× magnification. (f) Cells stained with Ki67 and Phalloidin at 20× magnification. (g) Cells stained with Ki67, DAPI, and Phalloidin at 40× magnification. (h-q) 2D and 3D heterogeneous assemblies of HUVEC, 3T3, and cardiomyocyte-encapsulating hydrogels. HUVECs, 3T3s, and cardiomyocytes are stained with Alexa 488 (green), DAPI (blue), and Propidium iodide (red), respectively. (h) Bright field and (i) fluorescence images of an assembly composed of circular and triangular gels. (j-o) Fluorescence images of several 2D heterogeneous assemblies of HUVEC, 3T3, and cardiomyocyte-encapsulating hydrogels. (p) Schematic form and (q) fluorescence image of 3D heterogeneous assembly of HUVEC, 3T3, and cardiomyocyte-encapsulating hydrogels. Scale bars are 500 μm unless otherwise is stated. 233

12.7 Frames from a movie with four teleoperated Mag-μMods assembling into a reconfigurable structure [61]. Copyright © 2011 by SAGE Publications, Ltd. Reprinted by permission of SAGE Publications, Ltd. Arrows indicate direction of magnetization. (a) Four Mag-μMods prepare for assembly. (b) All four modules are assembled in a T-configuration. (c) One module is broken free by rotation and reattaches in a new configuration. (d) The new assembly is mobile, and is shown moving to a new location. 239

List of Tables

Acknowledgments

This book could never be possible without the love and support of my beloved wife Seyhan and daughters, Ada and Doğa. They have made my life always more beautiful and meaningful. In addition to them, I have been so lucky and happy that my parents and two sisters have loved and supported me unconditionally all the time. My father has been a role model for me as a person and an intellectual with many ideals.

I created and taught my first Micro/Nano-Robotics course at UC Berkeley in 2002 as a lecturer with 43 PhD and 2 undergraduate students. It was an amazing first-time teaching experience, and I continued teaching it at Carnegie Mellon University for 11 years as a professor. The content of the course evolved and changed each time, and this book represents its latest version with a focus on mobile microrobotics mainly. I hope it will help any professor who wants to teach such course or anybody who wants to learn about or start working on microrobotics.

I have been lucky and privileged to conduct exciting high-impact research, mentor, and have fun socially with so many great post-docs and PhD, MSc, and undergraduate students since 2002. For this book, I specially thank my previous student Eric Diller, who wrote a tutorial on microrobotics with me in 2013, which formed a starting point for this book. I also thank my previous or current students or post-docs Chytra Pawashe, Steven Floyd, Rika Wright Carlsen, Jiang Zhuang, Slava Arabagi, Bahareh Behkam, Uyiosa Abusonwan, Burak Aksak, Bilsay Sümer, Çağdaş Önal, Michael Murphy, Yiğit Mengüç, Onur Özcan, Yun Seong Song, Zhou Ye, Joshua Giltinan, Hakan Ceylan, Ceren Garip, Lindsey Hines, Guo Zhan Lum, Xiaoguang Dong, and Shuhei Miyashita, whose works and papers have enabled some portions of the book. Moreover, I would like to thank Hakan Ceylan, Byungwook Park, and Ahmet Fatih Tabak for providing some text and references for a sub-section; Wendong Wang, Wenqi Yu, and Sukho Song for providing specific information for a table; and Lindsey Hines, Kirstin Petersen, Thomas Endlein, Massimo Mastrangeli, Byungwook Park, Rika Wright Carlsen, and Wendong Wang for reviewing and giving feedback for some chapters. Finally, I thank my assistant Janina Sieber for her help and Alejandro Posada for drawing many figures in this book.

1 Introduction

Significant progress in micro/nanoscale science and technology in last two decades has created increasing demand and hope for new microsystems for high-impact applications in healthcare, biotechnology, manufacturing, and mobile sensor networks. Such microsystems should be able to access small enclosed spaces such as inside the human body and microfluidic devices non-invasively and manipulate or interact with micro/nanoscale entities directly. Because human or macroscale robot sensing, precision, and size are not capable of achieving such desired characteristics, microrobotics has emerged as a new robotics field to extend our interaction and exploration capabilities to sub-millimeter scales. Moreover, mobile microrobots could be manufactured cost-effectively in large numbers, where a dense network of microrobots could enable new massively parallel, self-organizing, reconfigurable, swarm, or distributed systems. For these purposes, many groups have proposed various untethered mobile microrobotic systems in the past decade. Such untethered microrobots could enable many new applications, such as minimally invasive diagnosis and treatment inside the human body, biological studies or bioengineering applications inside microfluidic devices, desktop micromanufacturing, and mobile sensor networks for environmental and health monitoring.

1.1 Definition of Different Size Scale Miniature Mobile Robots

A typical macroscale mobile robot is a self-contained, untethered, and reprogrammable machine that can perceive, move, and learn in a given environment to realize a given task. But when can a mobile robot be called a mobile microrobot? Unfortunately, there is not yet a standardized definition of the term *microrobot*. Let us attempt to create a definition to classify different miniature robots in the literature. First, let us define two unique characteristics of a mobile microrobot [65]:

- *Overall size*: A mobile microrobot must be able to access small (less than 1 mm in all dimensions) spaces directly with minimal invasion, which entails untethered operation and all dimensions of the mobile robot being smaller than 1 mm.
- *Scaling effects on robot mechanics*: Locomotion mechanics and physical interactions of a mobile microrobot in a given environment are dominated by microscale physical forces and effects. Thus, volume-based forces such

as inertial forces, gravity, and buoyancy become almost negligible or comparable to surface area- and perimeter-based forces such as viscous forces, drag, friction, surface tension, and adhesion.

To incorporate these unique characteristics, we will define a mobile microrobot as *a mobile robotic system where its untethered mobile component has all dimensions less than 1 mm and larger than 1 μm and its mechanics is dominated by microscale physical forces and effects*. Thus, for microrobots, bulk forces are negligible or comparable to surface area- and perimeter-related forces. Also, viscous forces are much larger than inertial forces for a swimming microrobot, resulting in Reynolds number, which is the ratio of the inertial to viscous forces, less than 1. At the micron scale, fluid flows are mostly steady, and we are mostly in the Stokes flow regime. Brownian (stochastic) motion of microrobots in water resulting from their random collision with the water molecules at room temperature is negligible. Moreover, microrobots are made of sub-millimeter scale components, such as microactuators, microsensors, and micromechanisms, and are fabricated by microfabrication methods, which are different from conventional macroscale machining techniques. Finally, they have specific functions for a given task such as manipulation, sensing, cargo transport and delivery, and local heating.

There are currently two main approaches to designing, building, and controlling mobile microrobots in the literature depending on the given application:

- *On-board approach*: Similar to a typical macroscale mobile robot, the microrobot is self-contained and untethered, with all robot dimensions being less than 1 mm. Here, all on-board robot components, such as mechanisms, tools, actuators, sensors, power source, electronics, computation, and wireless communication, must be miniaturized down to few micrometers scale.
- *Off-board approach*: The mobile, untethered component of the microrobotic system is remotely (off-board) actuated, sensed, controlled, or powered and has all dimensions less than 1 mm while the overall system size could be very large.

The on-board approach is technically much more difficult to realize due to miniaturization challenges of all on-board components. However, it enables mobile microrobots navigating in large workspaces, e.g., in outdoors, which is required for mobile sensor network applications for environment monitoring and exploration. On the other hand, the off-board approach is easier to

Table 1.1

Definition of different size scale miniature mobile robots (Reynolds number is the ratio of inertial forces to viscous forces, which dictates the fluid dynamics regime.)

Mobile Robot Type	Overall Size	Dominant Forces Acting on Robot
Millirobots	1 mm to 10 cm	Macroscale volume-related forces; Reynolds number $\gg 1$
Microrobots	1 μm to 1 mm	Microscale surface area- or perimeter-related forces; Negligible Brownian motion; Reynolds number ~ 1 or $\ll 1$
Nanorobots	< 1 μm	Nanoscale physical and chemical forces; Non-negligible stochastic Brownian motion

implement due to fewer miniaturization challenges when operating in confined workspaces, such as the human body and microfluidic chips. Such limited workspace would not be an issue for potential microrobot applications in healthcare, bioengineering, microfluidics, and desktop micromanufacturing. Thus, almost all of the current mobile microrobotics studies in the literature have been using the off-board approach, and therefore our microrobotics definition also covers such studies.

In addition to the above on-board and off-board approaches, microrobots can also be classified as *synthetic* and *bio-hybrid*. In the former case, the microrobot is made of fully synthetic materials, such as polymers, magnetic materials, silicon, silicon oxide, metal alloys, composites, elastomers, and metals, while the latter is made of both biological and synthetic materials. Bio-hybrid microrobots are typically integrated with single or many cells, such as cardiac or skeletal muscle cells, or microorganisms, such as bacteria, algae, spermatozoids, and protozoa, and powered by the chemical energy inside the cell or in the environment. They harvest the efficient and robust propulsion, sensing, and control capabilities of biological cells at the microscale. Such cells could propel the robot in a given physiologically compatible environment, and sense environmental stimuli to control the robot motion by diverse mechanisms, such as chemotaxis, magnetotaxis, galvanotaxis, phototaxis, thermotaxis, and aerotaxis.

Reported miniature mobile robot sizes range from sub-micron to centimeter scale. We can classify such different length scale miniature robots as *millirobots*, *microrobots*, and *nanorobots* as given in Table 1.1. These small-scale robots have different dominant physical forces and effects. For the on-board approach case, their on-board components must have overall sizes much

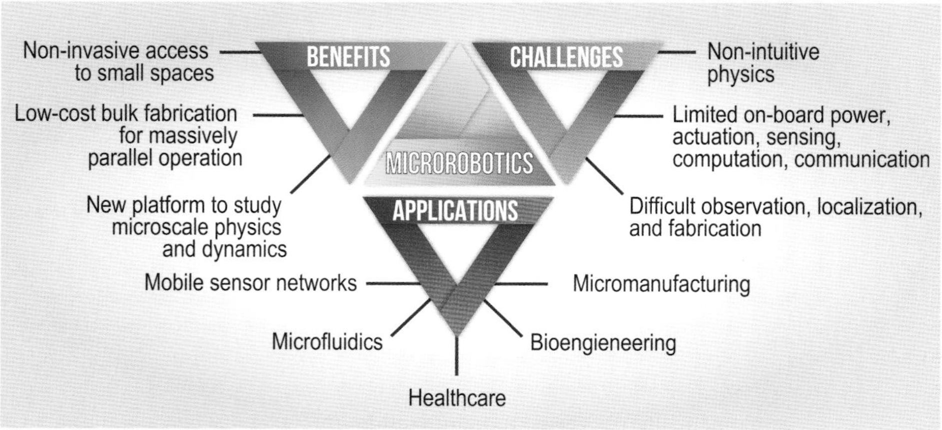

Figure 1.1
Diagram showing the benefits, challenges, and potential applications of mobile microrobots.

smaller than the given robot overall size. For millirobots, macroscale forces such as bulk forces dominate the robot mechanics instead of microscale forces and effects. The fluid dynamics is unsteady and even starts to be periodically turbulent when the Reynolds number is much larger than 1. For nanorobots, assumptions of continuum mechanics may not be valid at the sub-micron scale, and effects such as Brownian motion and chemical interactions create highly stochastic robot behavior. The fluid dynamics for nanorobots are no longer described accurately by the Navier-Stokes equation, so the Reynolds number is not relevant.

The size scale range in Table 1.1 presents significant new challenges in fabrication, actuation, locomotion mechanisms, and power supply not seen in macroscale mobile robotics. Microscale robots are particularly interesting because new physical principles begin to dominate the robot behavior. Changes in fluid mechanics, stochastic motions, and shorter time scales also challenge natural engineering notions as to how robotic elements move and interact. These physical effects must be taken into account when designing and operating robots at the small scale.

The benefits, challenges, and potential applications of mobile microrobots are overviewed in Figure 1.1. Here, we see that microrobots promise to access small spaces in a non-invasive manner as a new platform for microscale physics and dynamics. Compared with other robotic systems, they can be fabricated

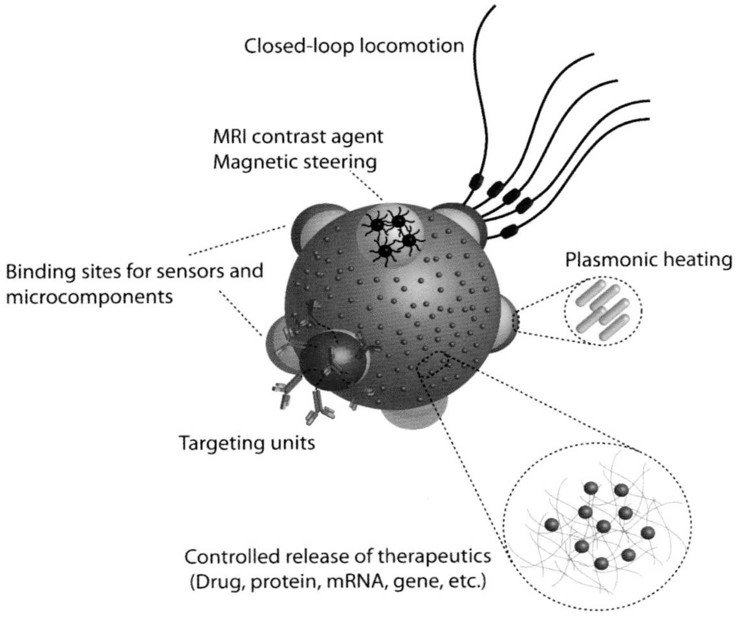

Figure 1.2

A conceptual sketch of an example future mobile microrobot with spatio-selective surface functionalization for potential medical applications. Each functional component could be assembled on a main body. The main body further could serve as a large depot for therapeutics to launch controlled release at the site of action. A closed-loop autonomous locomotion (e.g., a bio-hybrid design) could couple environmental signals to motility. Targeting units could enable reaching and localization at the intended body site. Medical imaging, e.g., magnetic resonance imaging (MRI), contrast agents loaded on the microrobot could enable visualization as well as remote steering on demand. Metallic nanorods could enable remote plasmonic or RF heating to decompose a tumor tissue by hypothermia.

inexpensively in bulk for potential massively parallel applications. However, several challenges arise in the design and control of microscale robots, such as non-intuitive attractive/repulsive and contact/non-contact physical forces, limited options for power and actuation, significant fabrication constraints, and difficulty in localizing such tiny robots. The field of microrobotics is particularly exciting due to the potential applications in healthcare, bioengineering, microfluidics, mobile sensor networks, and desktop microfactories. A conceptual sketch of an example mobile microrobot for medical applications is shown in Figure 1.2 with its possible components and functions.

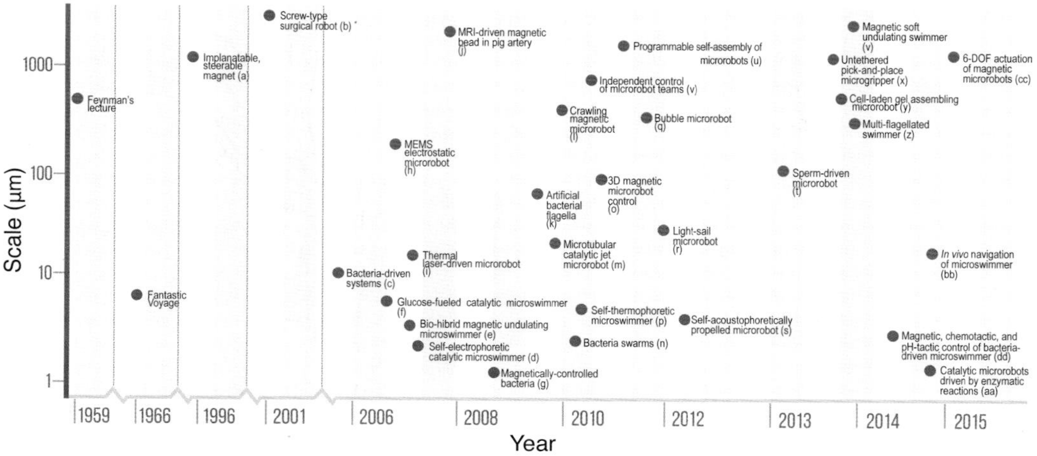

Figure 1.3

Approximate timeline showing the emerging new microrobot systems with their given overall size scale as significant milestones. (a) Implantable tiny permanent magnet steered by external electromagnetic coils [1]. (b) Screw-type surgical millirobot [2]. (c) Bacteria-driven bio-hybrid microrobots [3]. (d) Self-electrophoretic catalytic microswimmer [4]. (e) Bio-hybrid magnetic undulating microswimmer [5]. (f) Glucose-fueled catalytic microswimmer [6]. (g) Magnetically controlled bacteria [7]. (h) MEMS electrostatic microrobot [8]. (i) Thermal laser-driven microrobot [9]. (j) Magnetic bead driven by an MRI device in pig artery [10]. (k) Magnetic microswimmer with rigid helical flagellum inspired by bacterial flagella [11, 12]. (l) Crawling magnetic microrobot [13]. (m) Microtubular catalytic jet microrobot [14]. (n) Bacteria swarms as microrobotic manipulation systems [15]. (o) 3D magnetic microrobot control [16]. (p) Self-thermophoretic microswimmer [17]. (q) Bubble microrobot [18]. (r) Light-sail microrobot [19]. (s) Self-acoustophoretically propelled microrobot [20]. (t) Sperm-driven bio-hybrid microrobot [21]. (u) Magnetic, chemotactic, and pH-tactic control of bacteria-driven microswimmers [22–24]. (v) Magnetic soft undulating swimmer [25]. (x) Untethered pick-and-place microgripper [26]. (y) Cell-laden microgel assembling microrobot [27]. (z) Catalytic micromotors driven by enzymatic reactions [28]. (aa) In vivo navigation of microswimmers [29, 30]. (bb) 6-degrees-of-freedom (6-DOF) actuation of magnetic microrobots [31].

1.2 Brief History of Microrobotics

Advances in and increased use of microelectromechanical systems (MEMS) since the 1990s have driven the development of untethered microrobots. MEMS fabrication methods allow for precise features to be made from a wide range of materials, which can be useful for functionalized microrobots. There has been a surge in microrobotics work in the past few years, and the field is relatively new and growing fast [55, 66–68]. Figure 1.3 presents an overview of a few of the new microrobotic technologies which have been published, along with their approximate overall size scale.

The first miniature machines were conceived by Feynman in his lecture on "There's Plenty of Room at the Bottom" in 1959. In popular culture, the field of microrobotics is familiar to many due to the 1966 sci-fi movie Fantastic Voyage, and later the 1987 movie *Innerspace*. In these films, miniaturized submarine crews are injected inside the human body and perform non-invasive surgery. The first studies in untethered robots using principles which would develop into microrobot actuation principles were only made recently, such as a magnetic stereotaxis system [1] to guide a tiny permanent magnet inside the human body and a magnetically driven screw which moved through tissue [2]. Other significant milestone studies in untethered microrobotics include a study on bacteria-inspired swimming propulsion [69], bacteria-propelled beads [3, 70], steerable electrostatic crawling microrobots [8], laser-powered microwalkers [9], magnetic resonance imaging (MRI) device-driven magnetic beads [10], and magnetically driven milliscale nickel robots [71]. These first studies have been followed by other novel actuation methods, such as helical propulsion [11, 12], stick-slip crawling microrobots [13], magnetotactic bacteria swarms as microrobots [72], optically driven "bubble" microrobots [18], and microrobots driven directly by the transfer of momentum from a directed laser spot [19], among others. Figures 1.4 and 1.5 show a number of the existing approaches to microrobot mobility in the literature for motion in two dimensions and three dimensions. Most of these methods belong to the off-board (remote) microrobot actuation and control approach, and will be discussed in detail later. It is immediately clear that actual microrobots do not resemble the devices shrunk down in popular microrobotics depictions.

As an additional driving force for the development of mobile microrobots, the Mobile Microrobotics Competition began in 2007 as the "nanogram" league of the popular Robocup robot soccer competition [73]. This yearly event has since moved to the IEEE International Conference on Robotics and Automation and challenges teams to accomplish various mobility and manipulation tasks with an untethered microrobot smaller than 500 μm on a side. The competition has spurred several research groups to begin research in microrobotics, and has helped define the challenges most pressing to the microrobotics research field.

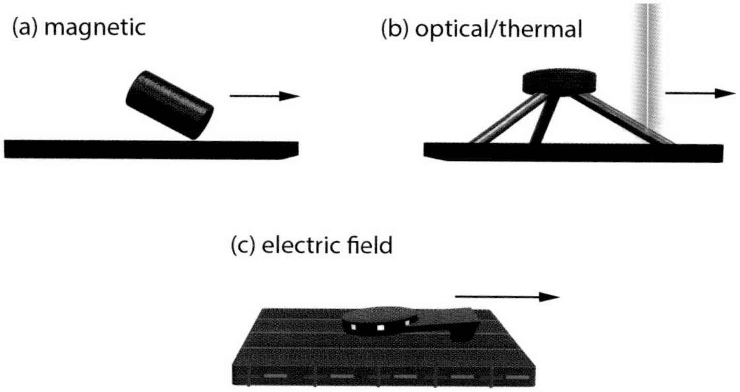

Figure 1.4
Some existing remote (off-board) approaches to mobile microrobot actuation and control in 2D. (a) Magnetically driven crawling robots include the Mag-μBot [13], the Mag-Mite magnetic crawling microrobot [32], the magnetic microtransporter [33], the rolling magnetic microrobot [34], the diamagnetically levitating milliscale robot [35], the self-assembled surface swimmer [36], and the magnetic thin-film microrobot [37]. (b) Thermally driven microrobots include the laser-activated crawling microrobot [9], the micro-light sailboat [19], and the optically controlled bubble microrobot [18]. (c) Electrically driven microrobots include the electrostatic scratch-drive microrobot [38] and the electrostatic microbiorobot [39]. Other microrobots which operate in 2D include the piezoelectric-magnetic microrobot MagPieR [40] and the electrowetting droplet microrobot [41].

1.3 Outline of the Book

This book introduces the reader to the newly emerging robotics field of mobile microrobotics. Chapter 2 covers the scaling laws that can be used to determine the dominant forces and effects at the micron scale. Such laws would also give us a significant physical intuition when we design and analyze different microrobots. Moreover, such scaling laws can be used to design and build scaled-up robots to understand the design and control principles for microrobotic systems, which are much harder to study experimentally at the micron scale directly.

In Chapter 3, forces acting on microrobots such as surface forces, adhesion, friction, and viscous drag are given and analytically modeled for simple spherical microrobot and flat surface interaction cases. Significant surface forces in air are typically van der Waals, capillary, and electrostatic forces for microsystems. In liquids, van der Waals forces still exist, but many other surface forces

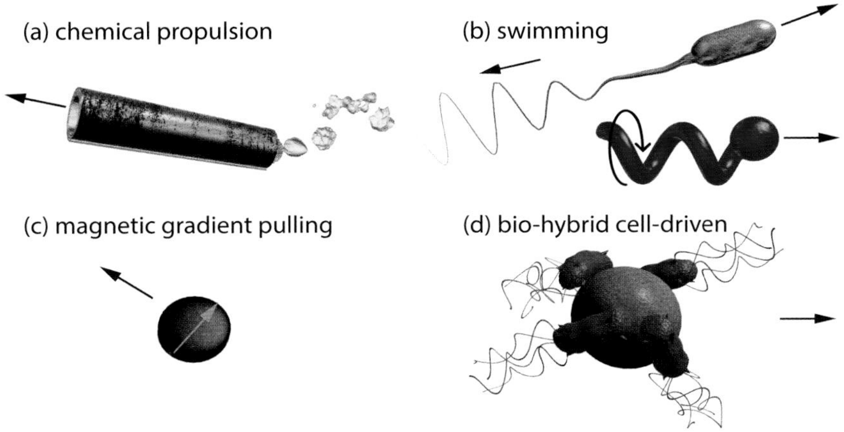

Figure 1.5

(a) Chemically propelled designs include microtubular jet microrobots [14], catalytic micro/-nanomotors [42], and electro-osmotic microswimmers [43]. (b) Swimming microrobots include the colloidal magnetic swimmer [5], the magnetic thin-film helical swimmer [44], the microscale magnetic helix fabricated by glancing angle deposition [12], the helical microrobot with cargo carrying cage fabricated by direct laser writing [45], and the helical microrobot with magnetic head fabricated as thin-film and rolled using residual stress [46]. (c) Microrobots pulled in 3D using magnetic field gradients include the nickel microrobot capable of 5-DOF motion in 3D using the OctoMag system [16] and the MRI-powered and imaged magnetic bead [47]. (d) Cell-actuated bio-hybrid approaches include the artificially magnetotactic bacteria [48], the cardiomyocyte-driven microswimmers [49], the chemotactic steering of bacteria-propelled microbeads [24], the sperm-driven and magnetically steered microrobots [21], and the magnetotactic bacteria swarm manipulating microscale bricks [15].

(such as double layer, hydration, and hydrophobic forces) also become important. When the microrobot contacts surfaces or other robots, surface forces induce adhesion, which is function of interfacial physical properties, contact geometry, and load. For elastic and viscoelastic materials, such adhesive forces and surface deformation are modeled using micro/nanoscale contact mechanics models. When the robot moves and inserts shear force on another solid surface it is in contact with, micro/nanoscale friction becomes crucial to model and understand. Sliding, rolling, and spinning types of frictional forces are modeled approximately. Inside fluids, microfluidic forces such as viscous drag and drag torque are important to model while having possible wall effects (i.e., changes in fluidic flows and forces due to the nearby walls) in the given operation environment. Finally, measurement techniques that can be used to characterize such micron scale force parameters are described so that the force models could use real empirical parameter values towards realistic robot behavior prediction.

Chapter 4 describes possible microfabrication techniques for microrobots, which are photo-lithography, bulk micromachining, surface micromachining, LIGA process, deep-reactive ion etching, laser micromachining, two-photon lithography, electro-discharge machining, micromilling, and so on. Each method's capabilities and limitations are studied so that the proper microfabrication method for a given microrobot design can be determined optimally. Especially, two-photon lithography is a recent exciting fabrication tool that could create a wide range of complex 3D microrobots with specific surface patterning and functionalization.

Chapter 5 includes possible on-board and remote sensing methods for microrobots. Tiny cameras and piezoresistive, capacitive, and piezoelectric microsensors could be potentially integrated to microrobots with proper size reduction, signal conditioning, and powering. However, such on-board sensors are not available for sub-mm scale robots, but remote magnetoelastic and optical sensing methods are more feasible for microrobots at the moment.

Microrobots can be actuated using on-board microactuators, self-propelled using physical or chemical interactions with their operation medium or biological cells attached to them, or remotely actuated. Chapter 6 studies possible on-board microactuators such as piezoelectric, shape memory alloy, conductive polymer, ionic polymer-metal composite, dielectric elastomer, MEMS electrostatic or thermal, and magneto- and electrorheological fluid actuators. Some of these actuators can be scaled down to micron scale as thin-film or unimorph/bimorph bending-type actuators integrated to robot structures directly while their on-board driving, control, and powering are still challenging for sub-mm scale robots. Chapter 7 describes self-propulsion methods that can use self-generated local gradients and fields or biological cells as the actuation source in proper liquid environments. Such catalytic (e.g., self-electrophoretic, self-diffusiophoretic, self-generated microbubbles-based, self-acoustophoretic, self-thermophoretic, and self-generated Marangoni flows based propulsion) or biological (bacteria, muscle cell, and algae-driven microswimmers) actuation approaches do not require any on-board electrical power source, electronics, processor, and control circuitry, which make

microrobot's locomotion mode, different external forces would be dominant. For swimming at the microscale, viscous forces would dominate the robot dynamics. A sphere with radius R moving in a liquid with speed U and a dynamic viscosity of μ would experience a viscous force of

$$F \approx 6\pi \mu RU, \tag{2.4}$$

which we could approximate as $F \approx \mu lU$. Using such a force model in Eq. [2.2] and $m = \rho l^3$, where ρ is the density:

$$\frac{mU^2}{Fl} = \frac{\rho l^3 U^2}{\mu \nu l^2} = \frac{lU}{\nu} = \text{Reynolds number}, \tag{2.5}$$

where $\nu = \mu/\rho$ is the kinematic viscosity. *Reynolds number* (Re) corresponds to the ratio of inertial forces to viscous forces. As the common fluids, dynamic (absolute) viscosities of air and water are around 0.02 cP and 1 cP, where 1 cP = 1.0 $\times 10^{-3}$ N.s/m^2, and kinematic viscosities of air and water are 15 cSt and 1 cSt, where 1 cSt = 1.0 $\times 10^{-6}$ m^2/s, respectively, at 20°C. As an example, Re = 10^{-4} for a swimming microrobot with a body length of 10 µm and an average speed of 10 µm/s in water, which means that viscous forces dominate its fluidic propulsion. As an important note here, l is the characteristic length in the motion direction that we like to calculate Re. For example, for calculating Re for a bacterium with an elliptic body shape of 2 µm length and 0.5 µm width swimming in the direction of its long length, l = 2 µm is selected.

As another example, for a flapping wing of a flying insect, $l = c$ where c is the mean wing chord length (maximum distance between the leading and trailing edges) and $U = 2f\phi_w l_w$ is the mean wing tip velocity where ϕ_w is the wing flapping amplitude in radians, f is the wing flapping frequency, and l_w is the wing length. So, for a fly with c = 2 mm, f = 200 Hz, and l_w = 8 mm, U = 9.6 m/s and Re = 1,200. If you like to build a scaled-up flapping system with c = 6 cm, f = 0.2 Hz, and l_w = 24 cm (U = 0.3 m/s) that has the same aerodynamics with a fly, we need to put such scaled-up system inside an oil tank where the oil viscosity should be around 13.5 times more viscous than water, i.e., ν = 13.5 cSt, to have both systems exhibit the similar aerodynamic behavior. Thus, this type of scaled-up system, such as in the case of RoboFly by Dickinson et al. [78], could enable detailed study of fly aeorodynamics using particle image velocimetry-type of fluidic measurement techniques, which is challenging to conduct at the real fly size scale.

Although gravitational force is typically not that significant for mobile microrobots conducting surface locomotion, such force could play a significant role for millirobots crawling, walking, or running on the ground. For such cases, $F = mg$ is the dominant force, where g is the gravitational acceleration constant, and

$$\frac{mU^2}{Fl} = \frac{mU^2}{mgl} = \frac{U^2}{gl} = \text{Froude number,} \tag{2.6}$$

where l is the characteristic length, e.g., the height of the hip joint from the ground in normal standing. *Froude number* (Fr) corresponds to the ratio of inertial forces to gravitational forces. Such a number is applicable to any legged ground locomotion robot or animal at the millimeter or larger scale. For example, all terrestrial biped animals, independent of their size scale, change their gaits from walking to running when Fr is larger than 0.5.

If the elastic forces dominate the robot mechanics, $F = k\Delta l$, where k is the stiffness of the elastic element and Δl is its deformation. Taking $\Delta l \propto l$, $F \approx kl$ and

$$\frac{mU^2}{Fl} = \frac{mU^2}{kl^2} = \frac{U^2}{\omega_0^2 l^2} = \text{constant} \Rightarrow \frac{fl}{U} = \text{Strouhal number,} \tag{2.7}$$

where $\omega_0^2 = k/m$ is the natural resonant frequency of the elastic system and f is the cyclic motion frequency. *Strouhal number* (St) corresponds to the ratio of added mass to inertial forces. Such a number is applicable to any microrobot having a cyclic motion regardless of whether elastic forces are dominant.

For microrobots locomoting at the interface of air (vapor) and water (liquid), surface tension dominates the microrobot dynamics. So, we can take $F = \gamma l$ and $m = \rho l^3$, where γ is the surface tension of the interface, which gives

$$\frac{mU^2}{Fl} = \frac{\rho l^3 U^2}{\gamma l^2} = \frac{\rho l U^2}{\gamma} = \text{Weber number.} \tag{2.8}$$

Weber number (We) corresponds to the ratio of inertial forces to surface tension. A water strider insect with a sub-millimeter scale leg diameter would have We $\ll 1$, meaning the repulsive surface tension would be the main lift mechanism, and the superhydrophobic (super water-repellent) wax-coated hairy insect legs would never penetrate the water surface. However, for a large animal such as a basilisk lizard running on water, We $\gg 1$, which means that the intertial forces would dominate the hydrodynamic lift of the lizard, and the lizard feet would penetrate the water surface, creating a lot of water splashes during water surface running [79].

Other than above non-dimensional numbers that are a function of inertial forces, some other useful numbers are relevant to mobile microrobots such as:

- *Bond Number* (Bo): Bond number corresponds to the ratio of buoyancy to surface tension, which is important for water surface locomotion robots. For microrobots, as in the case of water strider insects, Bo $\ll$ 1, which means that surface tension dominates buoyancy. It can be defined as

$$Bo = \frac{\rho g l^2}{\gamma}, \tag{2.9}$$

where ρ is the liquid density and l is the characteristic length, e.g., radius of a drop or the radius of a capillary tube.

- *Péclet Number* (Pe): Péclet number is defined as the ratio of transport caused by convection to transport by Brownian motion for mass diffusion, where

$$Pe = \frac{lU}{D}. \tag{2.10}$$

D is the mass diffusion constant. At small scales, Pe $<$ 1 means that transport is influenced by Brownian motion; in other words, small-scale objects are randomly jostled by colliding water molecules around them. Thus, robot dynamics would be dominantly stochastic if Pe $\ll$ 1, as in the case of nanoscale particles.

- *Capillary Number* (Ca): Capillary number is defined as the ratio of viscous forces to surface tension acting across an interface between a liquid and a gas, or between two immiscible liquids. It is defined as

$$Ca = \frac{\mu U}{\gamma}, \tag{2.11}$$

where γ is the surface or interfacial tension between the two fluid phases. For Ca $\gg$ 1, viscous forces dominate surface tension in the robot mechanics in the interface of water and air or two immiscible liquids.

2.2 Scaling of Surface Area and Volume and Its Implications

The dependence of a force on the characteristic scale L determines its relative influence at different size scales. The length dependence of some common forces in microrobotics is listed in Table 2.1. Scaling down an object's size with a length scale factor of L isomorphically (same geometric scaling

Table 2.1
Scaling of different forces dependent on length (perimeter), surface area, and volume

Perimeter Dependent ($\propto L$)	Surface tension
Area Dependent ($\propto L^2$)	Surface forces, Fluid drag, Friction, Reynolds number (Re), Evaporation, Fluid drag transient time τ (low Re), Heat conduction, Electrostatic forces
Volume Dependent ($\propto L^3$)	Mass, Inertia, Heat capacity, Buoyancy

in all three dimensions), its surface area and volume would scale down with L^2 and L^3, respectively. One major factor for many force balance comparisons is the surface area to volume ratio S/V, which is proportional to L^{-1}. As an example from the natural world, a meter-scale whale may have an S/V ratio of about 1 m^{-1}, while a micron-scale bacteria may have a S/V ratio of about 10^7 m^{-1}. This means that the surface area to volume ratio would increase at very small length scales such as micro- and nanoscale dimensions, and surface area-related forces and dynamics would start dominating at the microscale. For example, surface area-based powering methods such as solar cells would make more sense for milli- or microrobots while volume-based powering methods, such as batteries would provide almost negligible power to actuate robots smaller than 100 microns length scale. Moreover, small droplet evaporation gets much faster. In contrast, surface tension force, scaling with L^1, could dominate surface area and volume-related forces at the microscale. In a capillary tube, liquid surface tension dominates liquid weight such that 1-μm diameter hydrophilic capillary tube can raise water for around 30 m.

2.3 Scaling of Mechanical, Electrical, Magnetic, and Fluidic Systems

If it is assumed that the bulk values of material strength, modulus, density, friction coefficient, and so on are invariant with size, we can analyze how the behavior of mechanical mechanisms scales. The bending stiffness of a beam varies roughly with thickness^4/length^3 $\propto L^1$. However, deformation is proportional to force/stiffness (L^1), so the shape of a deformed structure is scale invariant. The resonant frequency of a vibrating system equals

$\sqrt{\text{stiffness/mass}} \propto L^{-1}$, so it increases as the size is reduced. An example resonant microscale robot which vibrates at a high frequency is the MagMite [32]. This two-mass system vibrates at several kHz frequency due to its small size of less than 200 μm. Mechanical power density (power per volume) scales with L^{-1}, which means high-power densities are possible for micron-scale robots.

Impact force scales with L^4 assuming acceleration scales with L^1, which means small-scale mobile robots are highly impact resistant. If you drop an ant or a tiny robot from tens or hundreds of meters high, it could still have no damage due to impact. Friction force scales with surface contact area (L^2) for adhesion dominated contacts at the microscale. Therefore, unlubricated rotary pin joints would easily wear out for microscale systems. So, using flexural joint designs instead of pin joints would be much more desirable and common for microscale systems.

Mechanical torque required to rotate a tiny mirror scales with L^5 where rotational mass moment of inertia scales with L^5 and rotational acceleration is assumed to scale with L^0. Thus, the required torque to rotate a tiny micromirror could be very small, meaning rotational actuation is much easier on the micron scale. For example, widely used digital light processing (DLP)-based vision display technology by Texas Instruments uses micromirrors that are rotated by electrostatic microactuators at very high speeds with very low torque requirement.

When we scale down the mechanical systems to nanoscale, above continuum mechanical force and parameter scaling relations are still valid while some corrections need to be made at the nanoscale. First, Young's modulus and strength of materials could change at the nanoscale, and we cannot assume they are constant any more. Single-wall carbon nanotube's modulus is about 1 TPa while bulk carbon's modulus is about 300 GPa. Also, metallic or other inorganic nanowires can have higher modulus and strength than their bulk materials due to reduced defects in nanoscale wires. Next, when quantum effects are dominant in a nanoscale system, there would be uncertainties in position and velocity/momentum.

In addition to the increase in resonant frequencies, relative locomotion speed (absolute locomotion speed normalized by the body length in the motion direction) of microscale organisms and robots increases and is much higher than large-scale animals and robots. Bacteria swim with speeds of 20-40 body lengths per second (blps), while a whale only swims at speeds of about 0.4 blps. In a similar way, microrobots can travel at speeds up to hundreds of blps, while large-scale robots typically travel up to several blps.

An attractive electrostatic force between two charged parallel plates scales by the square of the applied voltage, the square of the area of the plates, and the inverse square of the plate gap. For a constant voltage case, the elestrostatic force is $\propto L^0$, i.e., constant. To maximize the electrostatic force for stronger actuation force purposes, the voltage could be maximized close to the breakdown voltage, which scales with the gap size as $\propto L^1$. Thus, electrostatic force scales with L^2 in this case. However, at very small plate gap sizes, continuum assumption breaks down and the breakdown voltage scales with $L^{-1/2}$ (see empirical Paschen curves for an air gap [80]), which enables application of hundreds of Volts (V), e.g., up to 1,000 V in a few micron-scale gaps. In this non-continuum regime, electrostatic force scales with L^{-1}, meaning that electrostatic actuators with a few micrometers scale gaps could apply significant forces on the micron scale. Moreover, resistance of electrical wires is a function of *length/area*, which scales with L^{-1}, implying micron or smaller electrical lines would have high resistance and heating issues. For voltage $\propto L^1$, current scales with L^2, current density (current/area) is constant, i.e., $\propto L^0$, and capacitance $\propto L^1$.

Magnetic actuation forces could be exerted on a permanent microrobot remotely using a permanent magnet or an electromagnetic coil. A magnetic force between two permanent magnets is $\propto L^2$ because it scales as the square of the magnet volume and the inverse fourth power of the gap, while a magnetic torque between two permanent magnets is $\propto L^3$ because it scales as the square of the magnet volume and the inverse third power of the gap. However, isometric scaling may not be the best scaling method to compare these actuation methods because for a given actuation scheme, the gap or element size may be held constant. For example, in the case of a magnetic microrobot driven by external magnets, workspace limitations may mean that the actuation magnet may not come closer to the microrobot as the size scale decreases. For the case of an electromagnetic solenoid coil-based magnetic actuation, the magnetic force is $\propto I^2 R_c^2 / g^2$, where I is the current on the solenoid coil, R_c is the coil radius, and g is the distance between the coil and permanent magnet. Then, the magnetic force scaling depends on the current I scaling. Depending on the actuation system where the current density (current/area) is $\propto L^0$ (constant current density case), $\propto L^{-1/2}$ (constant heat flow case), and $\propto L^{-1}$ (constant temperature case), the force scaling would be $\propto L^4$, L^3, and L^2, respectively [81]. To maximize the magnetic fields, such as in the case of an MRI medical imaging system, a feedback-controlled liquid Helium (or water) cooling system around the coils (constant temperature case) is required. Thus, depending on the condition of the current density scaling, magnetic forces could be much smaller or comparable to the electrostatic forces on the micron scale. In general, however, it can be seen that electrostatic actuation forces scale down more favorably than magnetic ones, especially when operating at scales approaching 1 μm [82].

In microfluidic systems, Reynolds number Re scales with L^2 because it is the ratio of inertial ($\propto L^4$) to viscous ($\propto L^2$) forces. Such scaling law matches well with the biological data of swimming animals at different size scales. For microscale robots, Re is typically much less than 1, which implies a steady, laminar (Stokes) flow regime. In this regime, the fluid has a typically no-slip boundary condition with the microrobot and wall (e.g., microchannel) surfaces. It is much easier to model and analyze laminar flows compared with unsteady high Re flows. Fluid boundary layer thickness during the microrobot motion is at least several robot body lengths, which could induce fluidic forces on the neighbor objects or other robots. Due to such thick boundary layers, boundary conditions of the fluidic microsystem and fluidic coupling due to the distance among neigboring microrobots or surfaces/objects become more crucial. Moreover, in the Stokes regime, two liquids do not mix in the same microchannel, which requires mixing using external excitation. Pressure-based flow generation in a microchannel is almost impossible, and electrokinetic or surface tension-based flow generation is required.

If there is no active external fluidic mixing, mixing at the microscale could be mediated only by diffusion. Diffusion time of molecules in a fluid due to the Brownian motion scales with diffusion length^2 ($\propto L^2$) assuming the diffusion constant $D = k_B T/(6\pi \mu R)$ is constant [83]. Here, $k_B = 1.38 \times 10^{-23}$ $m^2 \cdot kg \cdot s^{-2} \cdot K^{-1}$ is the Boltzman constant, T is the temperature in Kelvin, R is the spherical molecule radius, and μ is the liquid dynamic viscosity. As an example, time to diffuse molecules over 10 μm is one million times faster than over 1 cm.

A summary of the scaling laws for overall relevant physical parameters and phenomena on the micron scale are listed in Table 2.2, for an isometric scaling law except where indicated.

2.4 Example Scaled-up Study of Small-Scale Locomotion Systems

Study of biologically inspired or artificial microrobotic locomotion systems is challenging because they are hard to visualize and fabricate, and they change

Table 2.2
Approximate isomorphic scaling relations and factors of different physical parameters

Physical Parameter	Scales With	Scaling Factor
Length	length	L^1
Surface area	length^2	L^2
Volume	length^3	L^3
Surface area/volume	area/volume	L^{-1}
Mass	volume	L^3
Bending stiffness	thickness^4/length^3	L^1
Beam resonant frequency	$\sqrt{\text{stiffness/mass}}$	L^{-1}
Deformation	force/stiffness	L^1
Elastic force	stiffness × deformation	L^2
Mechanical power	force × speed	L^2
Mechanical power density	power/volume	L^1
Microscale friction	contact area	L^2
Inertial or impact force	mass × acceleration	L^4
Mass moment of inertia	mass × length^2	L^5
Rotational torque	moment of inertia × acceleration	L^5
Surface tension	liquid contact perimeter	L^1
Breakdown voltage	gap size or gap size$^{-1/2}$	L^1 or $L^{-1/2}$
Electrostatic force	voltage^2 × area/gap^2	L^2, L^0, or L^{-1}
Capacitance	area/gap	L^1
Resistance	length/area	L^{-1}
Permanent magnet force	volume^2/gap^4	L^2
Current density	current/area	L^0, $L^{-1/2}$, or L^{-1}
Electromagnetic force	current^2 × radius^2/gap^2	L^4, L^3, or L^2
Reynolds number	inertial force/viscous force	L^2
Diffusion time	diffusion length^2	L^2

their physical design parameters. Therefore, scaled-up robot prototypes could help us understand the design and control principles for microscale robotic systems. As an example, it is hard to image, track, and change the flagella geometric parameters on a bacteria-inspired microswimmer using multiple helical flagella rotation to propel in liquids. An *E. coli* bacterium has an elliptic body with around 0.5-μm width and 2-μm length, and its multiple (typically 2-6) helical flagella has a diameter about 20 nm, helix amplitude about 200 nm, and length about 4 μm, which rotates at around 100 Hz. When each flagella rotates in the same direction at around 100 Hz, they bundle to each other and bacterium moves straightforward with an average swimming speed of 20 μm/s.

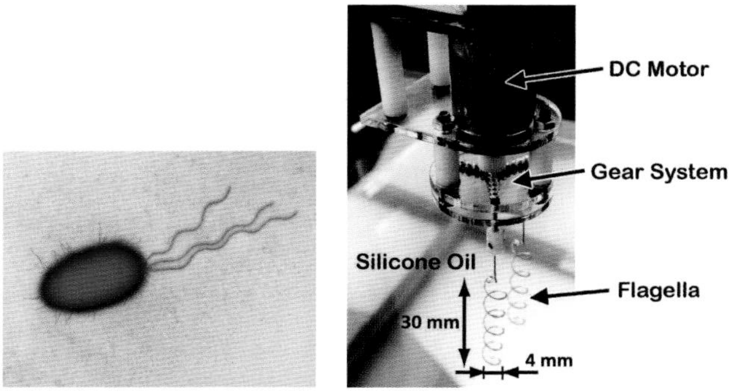

Figure 2.1

Left image: Transmission electron microscope image of a multi-flagellated *E. coli* bacterium with around 0.5-μm diameter and 2-μm length. Copyright © Dennis Kunkel Microscopy, Inc. Right image: Photograph of an example scaled-up multiple bacterial helical flagella setup for measuring the thrust force produced by multiple flagella as a function of flagella geometry and distance. A planetary gear system was used to guarantee individual rotation of each flagella at identical rates. Reprinted from [50] with the permission of AIP Publishing.

When any of the flagella rotation direction is changed at some random intervals at a given time, the flagella bundle opens up, and the bacteria tumbles to change its swimming direction. During the straight swimming of the bacterium in water, Re = 4×10^{-5} for the bacterium's straight body motion and Re = 2×10^{-3} for the bacterium's flagella rotational motion, which means that the viscous forces dominate its swimming dynamics.

To study the swimming propulsion and energy consumption speed effect of flagella length, helix amplitude, rotation frequency, and fluidic coupling among multiple flagella, we can build a scaled-up bacteria-like swimming robot prototype in centimeter scale. How could these two systems have the same swimming dynamics? As we mentioned in section 2.1, we need to have the same Re in two different size scale systems. For a large flagella model with an amplitude of 1 mm and a rotation frequency of 10 Hz, to have the same Re of 2×10^{-3}, we need to put the scaled-up model inside a highly viscous silicone oil with a kinematic viscosity 5×10^{4} times higher than water.

In [74], different flagella parameters were experimented using a scaled-up flagella system inside a silicone oil (Figure 2.1) to understand their effect on propulsion speed and efficiency. In [50], effect of multiple flagella spacing on bacteria propulsion force was studied using a scaled-up system. In this study, quantitative data showed that the reduced spacing among multiple flagella reduced the total propulsion force, which showed fluidic coupling among neighboring flagella. Also, after a critical rotation speed in the same rotation direction, the neighboring flagella started to attract and bend each other to create a bundle, which explains the fluidic force-based bundling of multiple flagella of the biological bacterium.

2.5 Homework

1. Calculate the Reynolds number for the below systems where $\nu = 10^{-6}$ m^2/s for water and $\nu = 1.5 \times 10^{-5}$ m^2/s for air:

   a. A bacterium with length 2 μm and width 0.5 μm swimming with speed of 20 μm/s.

   b. A moth flapping its wings (wing length: 3 cm; wing width [cord length]: 1 cm) at 50 Hz with an amplitude of 1.5 radians.

   c. To build a large-scale moth-inspired flapping wing-based flying robot with the flapping frequency of 0.5 Hz and amplitude of 1.5 radians dynamically similar to the above biological moth, calculate the kinematic viscosity of the oil that the robot should operate in.

   d. Compute the Strouhal number of the moth and the scaled-up moth robot above and compare their values. Should they be same?

2. Calculate how high water would rise in a glass capillary tube with a radius of 100 nm.

3. For a circular cross-section silicon beam fixed at its base with a length of 10 cm and radius of 2 mm, compute its first mode resonant frequency. Derive the scaling law for the resonant frequency of such beam, assuming an isometric length scaling. Using such a scaling law, compute the resonant frequency of the beam when you scale it down 10,000 times smaller.

4. Derive the scaling law for the electrostatic force between two parallel plates with length v, width w, gap distance d, and applied voltage V for the case of (a) gap distance of the plate above 15 μm, and (b) gap distance of the plate less than 2 μm. Also, derive the capacitance scaling law for such a plate.

5. Derive the scaling law for the magnetic force when a solenoid coil is used to create such a magnetic field for a constant (a) current density and (b) temperature.

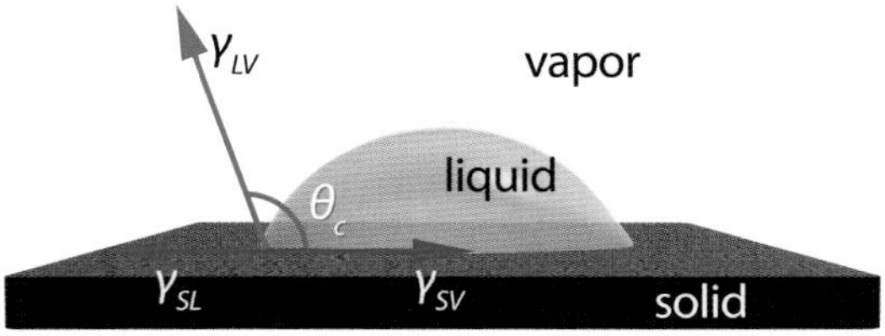

Figure 3.1
Schematics of surface tension forces acting at the liquid-solid-vapor interface, which result in a thermodynamic equilibrium contact angle θ_c on a solid surface in a given vapor such as air.

is in transition from fully hydrophilic to fully hydrophobic behavior. When $\theta_c > 150^\circ$, such surface is called *superhydrophobic*, where water droplets become almost like a spherical ball and roll off the surface easily. Contact angles of typical microrobot materials are listed in Table 3.2.

Water contact angle depends on the surface tension of the solid surface, which will be determined by the materials property or a chemical coating on the surface, and also on the surface roughness, i.e., geometry. To achieve a superhydrophobic surface, the hydrophobic surface material needs to have very low surface energy such as in the case of Teflon or it needs to be coated with a low-surface-energy mono- or multilayer materials such as PTFE. On the other hand, if the hydrophobic surface is also micro/nanostructured as in the case of lotus leaf surface, the water droplet contact angle could be increased up to $170^\circ - 180^\circ$. If the surface is hydrophilic, then micro/nanostructuring could further decrease the contact angle. To understand these geometric effects, we could analyze Wenzel and Cassie regimes as shown in Figure 3.2. If the non-structured atomically smooth surface has a static contact angle of θ_c and obeys to the condition of

$$\cos\theta_c < \frac{\phi_s - 1}{r_s - \phi_s}, \tag{3.2}$$

where ϕ_s is the total area fraction of solidliquid interface (≤ 1) and r_s is the roughness factor (= actual textured area / zero texture area ≥ 1), then the Cassie regime is valid and the new static contact angle θ_c^* on the micro/nanostructured or rough surface can be calculated as

$$\cos\theta_c^* = -1 + \phi_s(1 + \cos\theta_c); \tag{3.3}$$

Table 3.2

Surface tension (liquids) and energy (solids) and water contact angle in air for some common materials used in microrobotics

Material	**Surface Tension/ Energy** (mJ/m^2)	**Water Contact Angle** (°)
Water	72.8 [84]	-
Dodecane	25.4 [85]	-
Ethanol	22.0 [86]	-
Glycerol	64 [87]	-
Silicone oil	19.7 [88]	-
Eutectic gallium-indium (EGaIn)	624 [89]	-
Mercury	480 [89]	-
Iron	978 [90]	50 [91]
Iron oxide	1,357 [84]	$<$10 [92]
Pyrex Glass	31.5 [93]	39 [93]
Quartz	59.1 [94]	26.8 [94]
Silicon	1,250 [95]	35.7 [96]
Silicon dioxide	72 [97]	$<$10 [97]
Gold	1,283 [90]	~0 [98]
Silver	1,172 [90]	~0 [99]
Nickel	2,011 [90]	~6.2 [100]
Cobalt	2,775 [90]	~66 [101]
Platinum	2,299 [90]	~0 [98]
Copper	1,950 [90]	~7.4 [100]
SU-8	45.2 [102]	74 [102]
Polyurethane	37.5 [103]	77.5 [103]
Parylene-C	19.6 [104]	87 [104]
PDMS (polydimethylsiloxane)	21.8	107 [105]
PTFE (polytetrafluoroethylene)	19 [105]	109 [105]
PMMA (polymethyl methacrylate)	38 [105]	71 [105]
PEG (polyethylene glycol)	43 [105]	63 [105]
ABS (acrylonitrile butadiene styrene)	43 [105]	63 [105]
PET (polyethylene terephthalate)	39 [105]	72 [105]
PS (polystyrene)	34 [105]	87 [105]

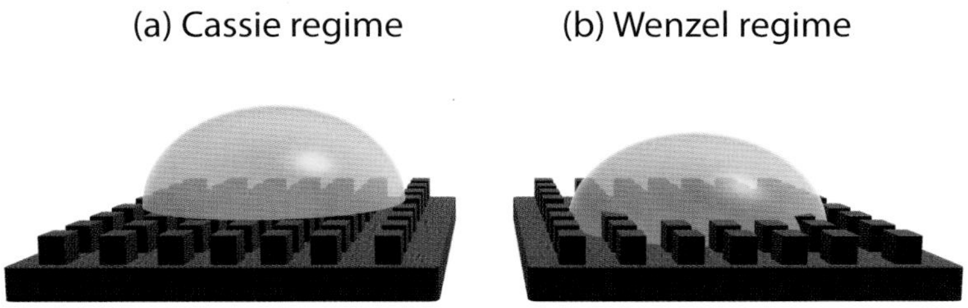

Figure 3.2
Schematics of liquid droplet wetting on structured surfaces: (a) Cassie regime where the liquid droplet does not fully wet the micro/nanostructure and there is air trapped under droplet, and (b) Wenzel regime where the droplet fully wets the micro/nanostructure.

otherwise, the Wenzel regime is valid and θ_c^* can be calculated as

$$\cos\theta_c^* = r\cos\theta_c. \tag{3.4}$$

In the Cassie regime, the water droplet stays on the micro/nanostructure, where air is trapped between the water-surface contact points under the droplet, as shown in Figure 3.2a. In the Wenzel regime, all areas under the water droplet are wetted fully and there is no air trapped, as shown in Figure 3.2b. Using such surface texture effects, superhydrophobic surfaces are possible for miniature robots that could repel water for lift generation, drag reduction (due to trapped air under surface texture), and other possible functions in air-liquid-solid interfaces.

3.2 Surface Forces in Air and Vacuum

Many small-scale mobile robots need to operate in ambient conditions in contact with or in close vicinity of surfaces and even in vacuum sometimes, such as inside a scanning electron microscope or a transmission electron microscope. In air and vacuum, microrobots experience surface forces such as van der Waals forces, electrostatic forces, hydrogen bonding, and capillary forces dominantly, as illustrated in Figure 3.3. In vacuum, capillary forces would not exist if the surfaces are fully dry. Magnetic forces could also exist in these conditions if there is any specific magnetic interaction among surfaces. We assume no special magnetic interaction between the robot and its operating surface in this chapter. Here, we will try to develop approximate continuum

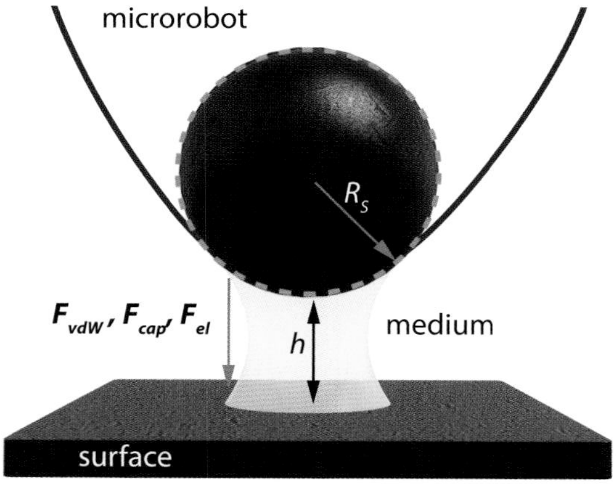

Figure 3.3

Short- or long-range, contact or non-contact, and attractive or repulsive surface forces, such as van der Waals, electrostatic, and capillary forces, create a sticky world for a microrobot interacting with other surfaces or objects in a given medium.

models of such surface forces assuming all surfaces are atomically smooth and have basic simple geometrical shapes.

3.2.1 van der Waals forces

van der Waals forces are polarization-type surface forces, which are due to instantaneous fluctuating dipole moments which act on nearby atoms within a material to generate an induced dipole moment. They have three components: dispersion, orientation, and induction forces. Dispersion (London) force acts on *all molecules*, and are quantum mechanical in origin. This force is called the dispersion force because the electronic movement that gives rise to it also causes dispersion of light, which is the variation of refractive index of a substance with the frequency (color) of the light. Orientation (Keesom) force exists between two surfaces with polar molecules, and it results from attraction between permanent dipoles. It is negligible for nonpolar molecules and inversely dependent on temperature. Induction (Debye) force exists between two surfaces with polar and non-polar molecules (i.e., permanent and induced dipoles), respectively. Debye force is independent of temperature.

London dispersion forces are typically very weak because the attractions are very quickly broken, and the charges involved are very small. The comparison of London forces with other bonding types is given in Table 3.1, which could give an idea of how relatively weak London forces are. Note that these comparison values are only approximate; the actual relative strengths will vary depending on the exact molecules involved.

van der Waals forces always exist, and are considered long-range surface forces, typically acting at ranges of 0.2 to 20 nm. The forces between two identical materials are always attractive but can be repulsive for certain cases involving different materials. Thus, the van der Waals forces are always important for microrobots in contact with objects or surfaces. For the case of an atomically smooth sphere-plane contact geometry, van der Waals forces can be approximately modeled as

$$F_{vdW}(h) \approx -\frac{A_{12}R}{6h^2}, \tag{3.5}$$

where R is the sphere radius, h is the separation distance, and A_{12} is the Hamaker constant between two surfaces. Here, it is assumed that $h \ll R$. The Hamaker constant is defined as

$$A_{12} = \pi^2 C \rho_1 \rho_2, \tag{3.6}$$

where C is the coefficient in the particle-particle pair interaction, and ρ_1 and ρ_2 are the number of atoms in the two interacting bodies per unit volume. Hamaker constants do not vary widely between different materials, and mostly lie in the range of (0.4-5) $\times 10^{-19}$ J = 40-500 zJ. Values for some common materials relevant to microrobots are given in Table 3.3.

The Hamaker constant for disparate materials with constants A_1 and A_2 in contact can be computed as

$$A_{12} \approx \sqrt{A_1 A_2}. \tag{3.7}$$

For two solids with constants A_1 and A_2 separated by a liquid with constant A_3, the Hamaker constant can be found as

$$A_{132} \approx \left(\sqrt{A_{11}} - \sqrt{A_{33}}\right)\left(\sqrt{A_{22}} - \sqrt{A_{33}}\right). \tag{3.8}$$

The van der Waals forces are reduced by the surface roughness because the two rigid surfaces are locally further separated. Taking b_r as the root-mean

Table 3.3

Hamaker constant for several common materials in vacuum, from [62] unless otherwise noted (1 zJ = 10^{-21} J)

Material	**Hamaker Constant** (zJ)
Air, vacuum	0
Water	37-40
Hydrocarbons	~50
Ethanol	42
Acetone	50
Glycerol	67
Silicone oil	~45
PS	65-79
PMMA	71
PDMS	45
Rubber	72
PTFE or Teflon	38
PMMA	63 [106]
Polymers	52-88 [106]
Mica	135
Graphite	275
Quartz	42-413 [106]
Fused quartz	65
Alumina	154
Silicon	221-256 [106]
Silicon dioxide	85-500 [106]
Silicon carbide	440
Gold	400
Silver	398
Copper	325
Iron	212 [106]
Iron oxide	210
Metals	300-500

square roughness of two surfaces separated by distance h, the force is attenuated as [107]

$$F_{vdW,\mathrm{rough}}(h) \approx \left(\frac{h}{h+b_r/2}\right)^2 F_{vdW}(h), \tag{3.9}$$

where h is measured to the top of the surface roughness. Typical roughness values are approximately 2 nm for a polished silicon wafer, and 1 μm for a polished metal surface.

Summary properties of van der Waals forces could be listed as:

- Dispersion component *always* exists for any surface interaction in any environment.
- They are maximum at full contact, i.e., when $h = a_0$ ($a_0 \approx 0.17$ nm typically).
- They are long range (effective at h = 0.2-20 nm typically).
- They are mostly attractive, but can be repulsive (depends on the sign of A_{132} in Eq. [3.8]).
- Non-additive (self-consistent solution possible using the method of charges).
- Retardation effects decline them rapidly after around 20 nm distance.
- Roughness significantly reduces them for rigid-rigid surface interactions while it may increase them for rigid-soft or soft-soft surface interactions.

3.2.1.1 Repulsive forces

The relation of van der Waals forces for very small separations can be investigated approximately using the Lennard-Jones potential, which is a simple model that predicts the change from attraction to repulsion at very small interatomic spacings due to the Pauli exclusion principle. The relation for two atoms separated by distance h is of the form

$$W(h) = -\frac{A_L}{h^6} + \frac{B_L}{h^{12}}, \tag{3.10}$$

where $W(h)$ is the interaction potential and $A_L = 10^{-77}$ J.m^6 and $B_L = 10^{-134}$ J.m^{12} depending on the type of atoms interacting [62]. Here, the first attractive term corresponds to the van der Waals energy. The minimum energy occurs at $h = 2^{1/6}\sigma = 1.12\sigma$, where σ is the atomic or molecular diameter. The interaction force using this potential can be computed as $F(h) = -dW(h)/dh$.

3.2.2 Capillary forces (surface tension)

The Young-Laplace equation relates the capillary pressure difference Δp to the shape of the surface sustained across the interface between two static fluids, such as water and air, due to surface tension. This equation is given as

$$\Delta p = \frac{\gamma}{r_k}, \tag{3.11}$$

where γ is the liquid surface tension and r_k is the radius of curvature.

Capillary forces act at fluid-air-solid interfaces to minimize the surface energy of the interface. As an example of the potential of these forces, they are used by trees to transport water from the roots to the leaves through capillaries several microns in diameter [108]. Thus, these forces can be relatively large in certain circumstances. In a capillary tube with circular cross-section (radius r_t), the interface between two fluids forms a meniscus that is a portion of the surface of a sphere with radius R, i.e., $r_k = R/2$. The pressure jump across this surface is:

$$\Delta p = \frac{2\gamma}{R}, \tag{3.12}$$

where R is a function only of the liquid-solid contact angle, θ_c, as

$$R = \frac{r_t}{\cos\theta_c}. \tag{3.13}$$

Thus,

$$\Delta p = \frac{2\gamma\,\cos\theta_c}{r_t}. \tag{3.14}$$

Then, the height h_c of liquid due to the capillary force is found using the Jurin equation as:

$$h_c = \frac{2\gamma\,\cos\theta_c}{\rho g r_t}, \tag{3.15}$$

where ρ is the fluid density and g is the acceleration due to gravity. As an example, for a water-filled glass tube of radius $r_t = 1$ μm, with $\gamma = 0.072$ N/m, $\theta_c = 20^\circ$, and $\rho = 1{,}000$ kg/m^3, the meniscus height will be 15 m.

The adhesive properties of surfaces in an ambient air environment can be sensitive to vapor in the environment due to capillary condensation, which forms in the cracks and pores of a surface [62]. The forces which occur when two objects come into contact in the presence of this thin liquid layer is dependent on the curvature of the spherical concave meniscus, which forms between. This curvature r_k can be found for the case of sphere-plane contact using the Kelvin equation [109] as:

$$r_k = \frac{\gamma\,V_l}{R_g T \log\left(p/p_s\right)}, \tag{3.16}$$

where V_l is the liquid volume, R_g is the gas constant, T is the temperature, and p/p_s is the relative humidity. For water at 20°C, $\gamma\,V/(R_g T) = 0.54$ nm, resulting in $r_k = 10$ nm at 90% humidity, $r_k = 1.6$ nm at 50% humidity, and $r_k = 0.5$

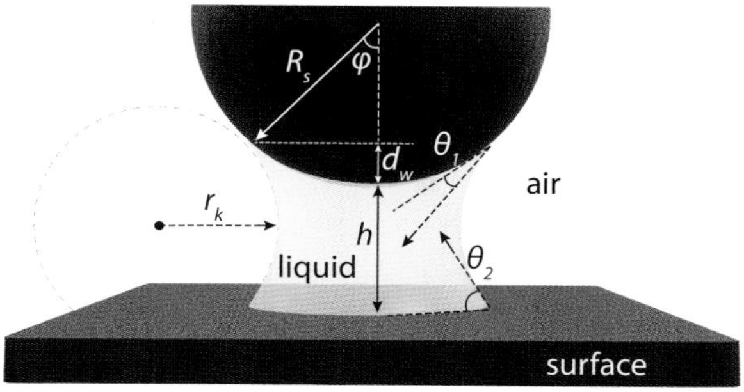

Figure 3.4
Schematics of parameters during a capillary force between a spherical microrobot asperity and a flat, smooth substrate due to the liquid bridge between two surfaces in air.

nm at 10% humidity. The resulting force can be derived approximately from the Laplace equation [62, 109] as

$$F_{cap} = -\frac{4\pi R\gamma \cos\theta_c}{1 + h/d_w}, \tag{3.17}$$

assuming $R \gg h$ and the liquid volume is very small, i.e., $\varphi \ll 1$, and constant. Here, R is the spherical asperity radius, θ_c corresponds to the effective liquid contact angle, where $2\cos\theta_c = \cos\theta_1 + \cos\theta_2$ (see Figure 3.4), h is the sphere-surface separation, and d_w is the immersion depth of the spherical asperity (see Figure 3.4). For a very small droplet ($\varphi \ll 1$) case, d_w can be computed as

$$d_w = 2r_k \cos\theta_c; \tag{3.18}$$

otherwise,

$$d_w = h\left(-1 + \sqrt{1 + \frac{V_l}{\pi R h^2}}\right). \tag{3.19}$$

If we do not like to have the assumption of $R \gg h$, the following model could be used:

$$F_{cap} = -\pi\gamma R \cos\theta_c \frac{(1 + \cos\varphi)^2}{\cos\varphi(1 + h/d_w)}. \tag{3.20}$$

The capillary force reaches to its maximum when the sphere is in contact with the surface and reduces as it is pulled away. For hydrophobic surfaces, i.e., $\theta_c > 90^\circ$, F_{cap} is repulsive. Capillary forces can immobilize microrobots such as the electrostatically driven scratch drive microrobot in [8]. This

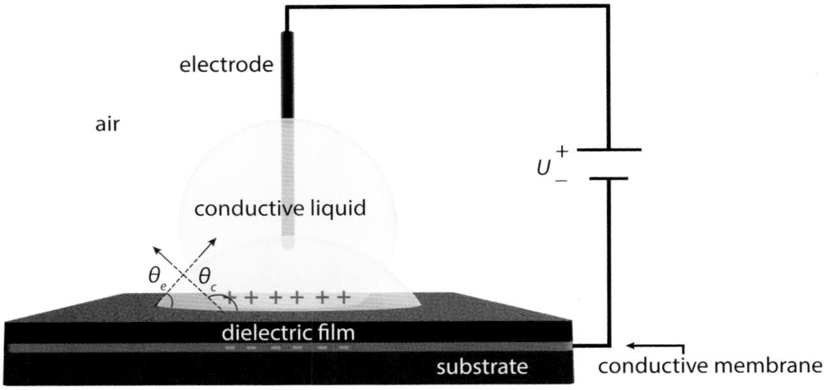

Figure 3.5
Schematics of electrowetting of a conductive liquid droplet on a flat, smooth dielectric film.

microrobot operates in a dry nitrogen environment with less than 15% relative humidity to prevent such effects. As another example, water strider insects with sub-millimeter diameter superhydrophobic legs repel water dominated by such repulsive capillary forces while buoyancy of their thin legs is negligible. In this way, they support their body weight using repulsive capillary forces [54, 110, 111].

3.2.2.1 Electrowetting

On a flat and smooth dielectric film, conducting fluids change their contact angle as a function of the applied DC voltage (see Figure 3.5). Lippmann-Young equation can be used to describe such angle change as

$$\cos\theta_e = \cos\theta_c + a_e, \tag{3.21}$$

where

$$a_e = \frac{\varepsilon\varepsilon_0 U^2}{2t_f\gamma}, \tag{3.22}$$

with t_f is the insulator film thickness, U is the applied voltage, and ε is the relative dielectric permittivity. Such a principle can be used for actively controlling the wetting of conductive liquids towards active focusing optical lenses or microactuation ideas for microrobotics.

3.2.3 Electrostatic forces

Electrostatic forces can develop between non-conductive objects resulting from charge buildup, electrical dipoles, or applied voltage [112]. Forces can also be induced between conductive surfaces when a voltage potential is applied to them. The force on a sphere of radius R at voltage potential U with a gap of distance h from a grounded plane is given by [113]

$$F_{el} = 2\pi\varepsilon\varepsilon_0 R^2 U^2 \left[\frac{1}{2(h+R)^2} - \frac{8R(R+h)}{(4(h+R)^2 - R^2)^2}\right], \tag{3.23}$$

where ε is the relative dielectric constant of the medium (e.g., $\varepsilon_{air} = \varepsilon_0$ and $\varepsilon_{water} = 80.4\varepsilon_0$) and $\varepsilon_0 = 8.85 \times 10^{-12}$ F/m is the dielectric constant of vacuum (vacuum permittivity). The relative permittivity of air is approximately 1, while that of water is 80.4. Electrostatic forces can be significantly smaller in water than in air because charge is able to dissipate through water. These electrostatic forces can be complex in practice and are often time-variant and difficult to measure. Therefore, in a microrobotics application, it is often best to reduce them. Electrostatic forces can be reduced by using conductive or grounded objects or by operating in conductive liquid environments. One common method when operating in an air or other non-conductive environment is to sputter a thin layer of conductive material onto the object surfaces and ground such a conductive layer to disperse the charges and reduce the electrostatic forces.

Electrostatic force between two different conductive surfaces is not zero even when no external voltage is applied. Due to the work function difference of different metals, electrostatic force still exists and is given by

$$F_{el} = -\frac{\pi\varepsilon_0 R}{h} U^2, \tag{3.24}$$

where $U = (\phi_1 - \phi_2)/e$ where ϕ_1 and ϕ_2 are the work functions of each metal and $e = 1.6 \times 10^{-19}$ C is the electron charge.

Electrostatic forces could be complex and hard to control for microrobots. For non-conducting microrobot materials, electrical charges could be created during surface contact motion due to triboelectrification in air and vacuum or they could already exist due to charges on surfaces beforehand. Such time-dependent or previously unknown charges could cause unexpected or time-dependent sticktion issues. Therefore, such surfaces should be coated with a thin metal layer (i.e., titanium-gold layer) and grounded if the application allows for preventing such uncontrolled and time-dependent charge accumulation. As the other option, the surfaces should be fully immersed under non-polar or non-conductive liquids to prevent such complex electrostatic forces. If the application allows only polar liquids, such as in the case of biological applications, then there would be new electrostatic forces in such polar liquids called double layer forces, which will be discussed later. However, double layer forces are much easier to control and not usually time-dependent.

3.2.4 Comparison of general forces on micron scale

A comparison of weight and general microscale surface forces is given in Figure 3.6. Here, typical values are chosen for a gold microsphere of radius R in close 0.2 nm contact with an infinite and atomically smooth gold plane for comparison. It can be seen that the sphere weight dominates at larger scales, but that van der Waals and capillary forces begin to come into play for objects smaller than several mm in size. Electrostatic forces only become dominant at much smaller sizes of several microns. These scaling relations can change if different materials are used, or if the surfaces are not atomically smooth, which would decrease the surface force magnitudes drastically.

In summary, when there is no external voltage applied and $R \gg h$, overall surface forces (F_T) between a conductive spherical and flat surface in air is:

$$F_T(h) \approx F_{vdW} + F_{cap} + F_{el} \approx -\frac{A_{12}R}{6h^2} - \frac{4\pi R\gamma \cos\theta_c}{1 + h/d_w} - \frac{\pi \varepsilon_0 R}{h} U^2, \qquad (3.25)$$

and when there is an external voltage applied:

$$F_T(h) \approx -\frac{A_{12}R}{6h^2} - \frac{4\pi R\gamma \cos\theta_c}{1 + h/d_w} + 2\pi \varepsilon\varepsilon_0 R^2 U^2 \left[\frac{1}{2(h+R)^2} - \frac{8R(R+h)}{(4(h+R)^2 - R^2)^2}\right]. \qquad (3.26)$$

3.2.5 Specific interaction forces

In addition to the previous surface forces, microrobotic surfaces could have specific interaction forces, such as Hydrogen bonding and magnetic and Casimir forces. Hydrogen bonding is possible for some microrobotics surfaces where such bond forms between a hydrogen atom that is bound to highly electronegative (electron-withdrawing) atoms such as O, N, or F, and an atom with an electron lone pair such as the oxygen in H_2O and R-OH, or the nitrogen

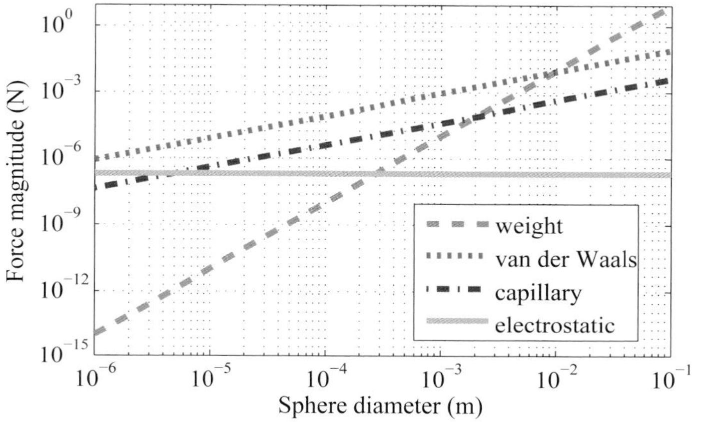

Figure 3.6

Comparison of weight and adhesive forces for a gold sphere in close 0.2 nm contact with a gold surface. The surfaces are assumed atomically smooth gold, with a Hamaker constant of 400 zJ for gold. The medium is assumed as air, with a relative permittivity of 1. The sphere was assumed to be made from gold ($\rho = 19,300$ kg/m^3) for the weight calculation, and to have a voltage of 100 V for the electrostatic calculation in an air environment. For the capillary force, $\gamma = 0.0728$ N/m is chosen for water with $\theta_c = 85°$ for water contact angle on gold.

in H_3N or R_3N. They are responsible for the large surface tension of water (compared with molecules of similar size), the adsorption of water molecule on metal oxide surfaces, and the binding of DNA double helices. Magnetic forces are possible when the microrobot body and surfaces they interact with contain magnetic materials. The Casimir effect can occur if the microrobot and the surface it is interacting with are uncharged, conducting, atomically smooth, and parallel flat plates in vacuum. Due to quantum vacuum fluctuations of the electromagnetic field, a tiny attractive Casimir force F_{cas} between two plates of area A_c separated by a distance h would be generated, which can be given by [114],

$$F_{cas}(h) = \frac{\pi \hbar c_l}{480 h^4} A_c, \tag{3.27}$$

where $\hbar = 6.6 \times 10^{-34}$ is the Planck's constant and $c_l = 3 \times 10^8$ m/s is the speed of light.

3.2.6 Other geometries

Force relations for other contact geometries are more complex than a simple sphere-plane or plane-plane case, where we need to numerically compute surface forces for complicated 3D surface geometries. However, for sphere-sphere and cylinder-cylinder type of specific surface interactions, *Derjaguin approximations* are very useful to compute surface force models using the plane-plane interaction models for any force law. Here, we assume $h \ll R$, where R is the radius of the sphere or cylinder. For a known plane-plane of interaction potential model of $W(h)_{plane-plane}$, here are Derjaguin approximations to drive other geometries [62]:

$$F_{sphere-sphere}(h) \approx 2\pi \left(\frac{R_1 R_2}{R_1 + R_2} \right) W_{plane-plane}(h), \tag{3.28}$$

$$F_{cylinder-cylinder}(h) \approx 2\pi \left(\frac{\sqrt{R_1 R_2}}{sin\theta_s} \right) W_{plane-plane}(h), \tag{3.29}$$

where R_i is the radii of two spheres or cylinders ($i = 1, 2$) and θ_s is the angle between two horizontally crossed cylinders. If $R_2 \gg R_1$, as in the case of a sphere-plane case, $F_{sphere-plane}(h) \approx 2\pi R_1 W_{plane-plane}(h)$ can be computed.

For example, for van der Waals forces,

$$W_{plane-plane}(h) = -\frac{A_{12}}{12\pi h^2}, \tag{3.30}$$

which results in

$$F_{sphere-sphere}(h) \approx -\frac{A_{12} R_1 R_2}{6(R_1 + R_2)h^2}, \tag{3.31}$$

$$F_{sphere-plane}(h) \approx -\frac{A_{12} R}{6h^2}, \tag{3.32}$$

where $R = R_1 \ll R_2$. Moreover,

$$F_{cylinder-cylinder}(h) \approx -\frac{A_{12}\sqrt{R_1 R_2}}{6h^2} \tag{3.33}$$

can be computed immediately as another basic contact geometry assuming h is much smaller than R, R_1, and R_2. For a two perpendicular and parallel cylinders, these approximations do not apply, and we compute it from continuum approximations as [62]:

$$F_{cylinder-cylinder}(h) = \frac{-A_{12} l_c}{8\sqrt{2} h^{5/2}} \left(\frac{R_1 R_2}{R_1 + R_2} \right), \tag{3.34}$$

where l_c is the cylinder length.

3.3 Surface Forces in Liquids

Many microrobots are required to operate in liquid media for biomedical, some micromanufacturing, and microfluidic applications. Therefore, it is crucial to know how surface forces change due to liquid medium to be able to design the proper robot materials, shape, and size.

3.3.1 van der Waals forces in liquids

van der Waals forces exist in every environment with the given models as we mentioned in Section 3.2.1. In liquids, as the only change, the effective Hamaker constant (A_{132}) becomes much smaller for typical liquids. For example, for a polystyrene robot body interacting with a gold-coated surface under deionized (DI) water, the effective $A_{PS-water-Au}$ is around 28 zJ (calculated from Eq. [3.8]) while for the same materials the effective $A_{PS-air-Au}$ is 162.5 zJ. There is more than a five times drop in the van der Waals forces for this case. If we immerse the same two materials under silicone oil, then $A_{PS-oil-Au}$ would be 18.8 zJ.

3.3.2 Double-layer forces

In polar liquids, the typical electrostatic forces are double-layer forces. Every material has induced surface charges on their surfaces in contact with a highly polar liquid such as water by dissociation of ions from the surfaces into the solution or preferential adsorption of certain ions from the solution. Most typical materials have a negative surface charge under water while some biological or synthetic materials, such as polyethylenimine (PEI), poly-L-lysine (PLL), poly(allylamine hydrochloride) (PAH), and chitosan, have a positive surface charge under water. The surfaces maintain their electrical neutrality by an opposite charge layer at a small distance where such charge layer thickness is called *Debye length* (κ). Ionic concentration, i.e., salt concentration in moles (M), of the polar liquid medium plays a significant role for κ, where

$$\kappa^{-1} = \frac{0.304}{\sqrt{M_{11}}} \text{ for 1:1: electrolytes such as NaCl,} \tag{3.35}$$

$$\kappa^{-1} = \frac{0.174}{\sqrt{M_{12}}} \text{ for 1:2 or 2:1 electrolytes such as CaCl}_2, \tag{3.36}$$

$$\kappa^{-1} = \frac{0.152}{\sqrt{M_{22}}} \text{ for 2:2 electrolytes such as MgSO}_4. \tag{3.37}$$

As an approximate model for the double layer force (F_{DL}) for two flat and atomically smooth planes,

$$F_{DL}(h) = \frac{2\zeta_1\zeta_2}{\varepsilon_e\varepsilon_0} A_c e^{-h/\kappa}, \tag{3.38}$$

where ζ_1 and ζ_2 are the surface charge densities (in other words, zeta potentials) of the two surfaces with the surface area A_c and ε and $\varepsilon_0 \approx 8.85 \times 10^{-12}$ F/m are the liquid and vacuum dielectric constants, respectively. Such long-range force is repulsive for same-surface charge materials and attractive for opposite-surface charge materials.

3.3.2.1 DLVO theory

Derjaguin-Landau-Verwey-Overbeek (DLVO) theory from 1945 combines van der Waals and double-layer forces in polar liquids as:

$$F_{DLVO}(h) = \frac{2\zeta_1\zeta_2}{\varepsilon\varepsilon_0} A_c e^{-h/\kappa} - \frac{A_{132}}{6\pi h^3} A_c. \tag{3.39}$$

3.3.3 Hydration (steric) forces

If two *hydrophilic* surfaces are immersed in polar electrolytes, then long-range hydration forces repel each other with the energy of

$$E_{Hphil}(h) = E_0 e^{-h/\lambda_0}, \tag{3.40}$$

where $\lambda_0 \approx$ 0.6-1.1 nm and E_0 = 3-30 mJ/m^2. Thus, the resulting repulsive force is:

$$F_{Hphil}(h) = -\frac{dE_{Hphil}}{dh} = \frac{E_0}{\lambda_0} e^{-h/\lambda_0}. \tag{3.41}$$

3.3.4 Hydrophobic forces

If two *hydrophobic* surfaces are immersed in polar electrolytes, then long-range hydrophobic forces attract each other with the energy of

$$E_{Hphob}(h) = -2\gamma\, e^{-h/\lambda_0}, \tag{3.42}$$

where $\lambda_0 \approx$ 1-2 nm and γ = 10-50 mJ/m^2. Thus, the resulting force is:

$$F_{Hphob}(h) = -\frac{dE_{Hphob}}{dh} = -\frac{2\gamma}{\lambda_0} e^{-h/\lambda_0}. \tag{3.43}$$

3.3.5 Summary

In polar liquids, the total surface forces (F_T) between two flat, smooth surfaces is:

$$F_T(h) \approx F_{DLVO} + F_{Hphil} = \frac{2\zeta_1\zeta_2}{\varepsilon\varepsilon_0} A_c e^{-h/\kappa} - \frac{A}{6\pi h^3} A_c + \frac{E_0}{\lambda_0} e^{-h/\lambda_0} \tag{3.44}$$

for hydrophilic surfaces and is:

$$F_T(h) \approx F_{DLVO} + F_{Hphob} = \frac{2\zeta_1\zeta_2}{\varepsilon\varepsilon_0} A_c e^{-h/\kappa} - \frac{A}{6\pi h^3} A_c - \frac{2\gamma}{\lambda_0} e^{-h/\lambda_0} \tag{3.45}$$

for hydrophobic surfaces. In non-polar liquids with no other specific interactions:

$$F_T(h) \approx -\frac{A_{132}}{6\pi h^3} A_c. \tag{3.46}$$

As the summary properties of surface forces in liquids:

- Forces in liquids are generally much less than air and vacuum;
- Hydrophobic surfaces attract and hydrophilic ones repel in polar liquids;
- Liquid medium parameters such as salt concentration and temperature can change the force values and behavior significantly;
- If allowable, non-polar liquids would be the best medium to design for mobile microrobot applications with less complex surface forces.

3.4 Adhesion

Surface forces in the previous section usually create attractive forces for microrobots when they *contact* to a surface or object; therefore, we need to apply force and energy to separate the robot from the surface or object. The maximum force needed to separate two such bodies is called *adhesion* (pull-off force, P_{po}). Moreover, the required maximum energy per unit area to separate two surfaces is called *work of adhesion* (W_{12}). If these two surfaces have the same material, $W_{12} = 2\gamma$, where γ is the surface energy of the given surface material. For two different surfaces with intrinsic surface energies of γ_1 and γ_2 [62, 112],

$$W_{12} = \gamma_1 + \gamma_2 - \gamma_{12} \approx \sqrt{2\gamma_1^d \gamma_2^d} + \sqrt{2\gamma_1^p \gamma_2^p}, \tag{3.47}$$

where γ_{12} is the interfacial surface energy between two surfaces and $\gamma = \gamma^d + \gamma^p$ with γ^d and γ^p corresponding to the dispersive and polar components of the surface energy, respectively.

If the medium is not vacuum or air, then the medium also contributes to the effective work of adhesion between two surfaces as the third material so that

$$W_{132} = W_{12} + W_{33} - W_{13} - W_{23}. \tag{3.48}$$

Positive or negative total work of adhesion values W_{132} can result from different material and liquid layer combinations. Negative values imply that the two surfaces repel each other, whereby the surfaces minimize their energy by contacting the fluid, not each other. As an example of this case, a glass surface was used with polystyrene beads for manipulation in [115]. The range of immersed work of adhesion when immersed in water would be $-45\ \text{mJ}\cdot\text{m}^{-2} < W_{132} < -3.1\ \text{mJ}\cdot\text{m}^{-2}$. Because this range is necessarily negative, these two surfaces would repel each other. This repulsion can aid in object motion, as the motion will be governed primarily by fluid interactions. For irregularly shaped particles, the surface forces will also be greatly reduced from its perfectly smooth value.

For inert non-polar surfaces, e.g., hydrocarbons, adhesion is just due to the dispersion van der Waals forces, i.e., $\gamma = \gamma^d$, and work of adhesion would be due to the van der Waals energy between two flat surfaces W_{planes} in contact as:

$$W_{12} = W_{planes}(a_0) = \frac{A_{12}}{12\pi a_0^2}, \tag{3.49}$$

where a_0 is the interatomic distance. If the two interacting surfaces are the same, then:

$$\gamma = \frac{A}{24\pi a_0^2}. \tag{3.50}$$

For $a_0 \approx 0.165$ nm and a nonpolar surface with $A = 50$ zJ case, $\gamma = 24.4\ \text{mJ/m}^2$ can be estimated from this relation, or vice versa.

3.5 Elastic Contact Micro/Nanomechanics Models

Many continuum contact mechanics theories exist, starting from Hertz in 1882 to Johnson-Kendall-Roberts (JKR) theory in 1971, Derjaguin-Muller-Toporov (DMT) theory in 1975, and Maugis-Dugdale (MD) theory in 1992 [116–120].

Let us understand first each of these theories for the case of a spherical object in contact with a flat surface and see which one or ones can be used to model the contact deformation and adhesion of a given microrobot case. All of the below contact mechanics models assume:

- An atomically smooth sphere is in contact with an atomically smooth flat surface;
- The bodies in contact are elastic;
- There are no viscoelastic effects on the bodies and at the interface;
- Contact radius of the spherical body on the flat surface is much smaller than the radius of the spherical body; and
- There is no friction and wear at the interface.

In the Hertz theory, only an external load compresses and deforms the interface with the pressure profile shown in Figure 3.7(a) given as

$$p(r) = p_0\sqrt{1-(r/a)^2}, \quad r < a, \tag{3.51}$$

which produces a parabolic deformation curve. Such pressure profile deforms in the interface with below relations:

$$a = \left(\frac{RL}{K}\right)^{1/3}, \tag{3.52}$$

$$\delta = \frac{a^2}{R}, \tag{3.53}$$

$$p_0 = \frac{3Ka}{2\pi R}, \tag{3.54}$$

$$K = \frac{4}{3}\left(\frac{1-\nu_1^2}{E_1} + \frac{1-\nu_2^2}{E_2}\right)^{-1}, \tag{3.55}$$

where a is the contact radius of the spherical body on the flat surface (Figures 3.7(a) and 3.8(a)), δ is the indentation depth, L is the external normal load, R is the spherical body radius, and K is the equivalent elastic modulus of contact based on each material's Young's modulus (E_1 and E_2) and Poisson's ratio (ν_1 and ν_2) [62]. For this model, if $L = 0$, then $a = 0$, and there would be no deformation.

In the DMT theory, in addition to the contact deformation due to the load as given by the Hertz model, a long-range adhesive stress outside of the contact

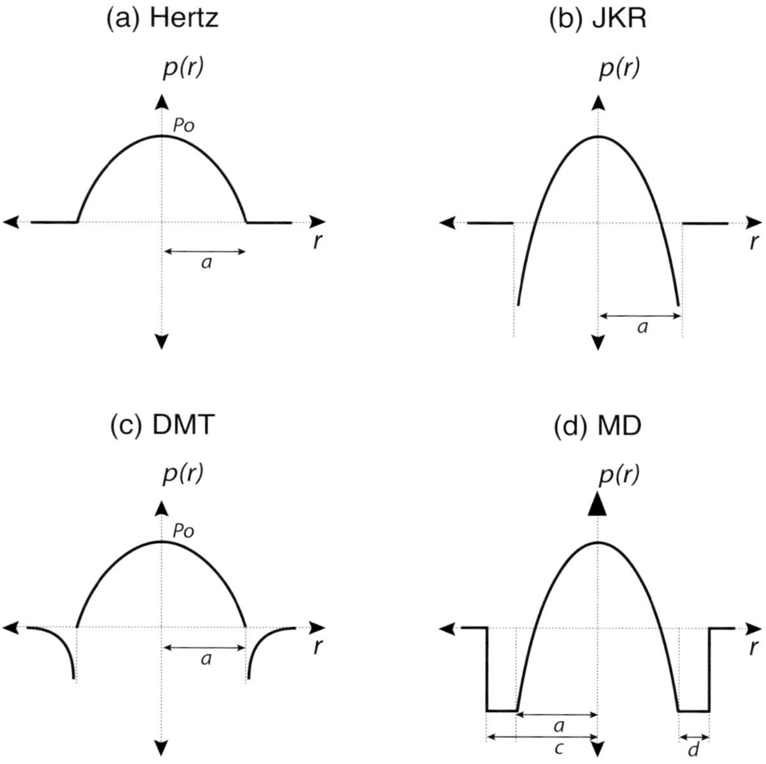

Figure 3.7

Pressure (stress) profiles for (a) Hertz, (b) JKR, (c) DMT, and (d) MD elastic contact mechanics models, where compressive (positive sign) external normal load stress and tensile (negative sign) adhesive stress create small deformation on a spherical object in contact with a flat plane.

area (Figures 3.7(c) and 3.8(a)) is introduced. Then,

$$a = \left(\frac{R}{K}(L + 2\pi RW_{12})\right)^{1/3}, \tag{3.56}$$

$$\delta = \frac{a^2}{R}, \tag{3.57}$$

$$P_{po} = -2\pi RW_{12}, \tag{3.58}$$

where P_{po} is adherence (pull-off) force between the sphere and the plane. Here, if $L = 0$, then there would still be a finite contact radius due to surface forces.

In the JKR theory, a short-range adhesive stress inside the contact area (Figures 3.7(b) and 3.8(a)) is introduced while having an infinite stress at the periphery. Then,

$$a = \left(\frac{R}{K}\left[L + 3\pi RW_{12} + \sqrt{6\pi RW_{12}L + (3\pi RW_{12})^2}\right]\right)^{1/3}, \quad (3.59)$$

$$\delta = \frac{a^2}{R} - \frac{8\pi aW_{12}}{3K}, \quad (3.60)$$

$$P_{po} = -1.5\pi RW_{12}. \quad (3.61)$$

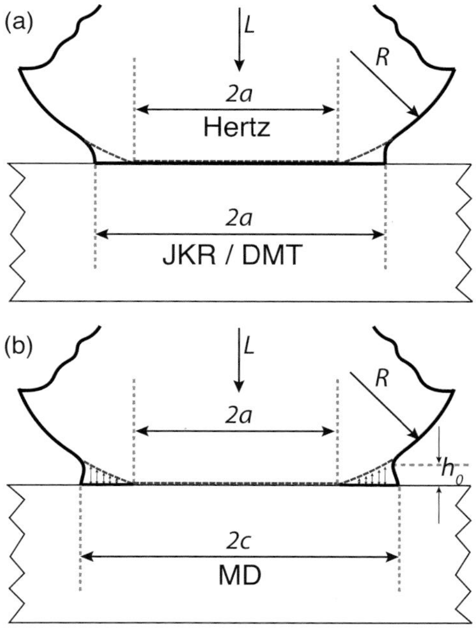

Figure 3.8

Contact geometry of a sphere deforming on a flat plane by (a) Hertz, JKR, and DMT models; and (b) MD elastic contact mechanics model for an applied external load of L and attractive surface forces increasing the contact deformation differently depends on the contact model. The MD theory models the periphery of the sphere-plane interface as a crack.

As the most recent, general, and accurate model, the MD theory models the periphery of sphere-plane interface as a crack using fracture mechanics models [121, 122] (Figures 3.7(d) and 3.8(b)). Assuming that a constant attractive stress (modeled as the theoretical adhesive strength of the interface that is approximated by the Lennard-Jones potential as $\sigma_0 = W_{12}/h_0 = 1.03W_{12}/a_0$) acts over a cohesive zone distance d, then:

$$L = \frac{Ka^3}{R} - \lambda a^2 \left(\frac{\pi W_{12}}{RK}\right)^{1/3} \left[\sqrt{m^2-1} + m^2 arctan\sqrt{m^2-1}\right], \quad (3.62)$$

$$1 = \frac{\lambda a^2}{2}\left(\frac{K}{\pi R^2 W_{12}}\right)^{2/3}\left[\sqrt{m^2-1} + (m^2-2) arctan\sqrt{m^2-1}\right]$$
$$+\frac{4\lambda^2 a}{3}\left(\frac{K}{\pi R^2 W_{12}}\right)^{1/3}\left[1 - m + \sqrt{m^2-1} arctan\sqrt{m^2-1}\right] \quad (3.63)$$

$$\delta = \frac{a^2}{R} - \frac{4\lambda a}{3}\left(\frac{\pi W_{12}}{RK}\right)^{1/3}\sqrt{m^2-1}, \quad (3.64)$$

$$\alpha = \frac{7}{4} - \frac{1}{4}\frac{4.04\lambda^{1/4}-1}{4.04\lambda^{1/4}+1}, \quad (3.65)$$

$$P_{po} = -\alpha\pi R W_{12}. \quad (3.66)$$

Here, $1.5 \le \alpha \le 2$, $m = c/a$, $c = a + d$, and a_0 is the interatomic distance. After solving the first two closed-form equations for a and m for a given L and other system parameters, δ can be computed. Moreover, λ is the dimensionless Tabor parameter, which is defined as:

$$\lambda = \frac{2.06}{a_0}\left(\frac{W_{12}^2 R}{\pi K^2}\right)^{\frac{1}{3}}. \quad (3.67)$$

For a given microrobot in contact with a flat surface (assuming a spherical asperity contact for the microrobot with the surface), which above contact mechanics model should we use? First, if the external load on the robot is too high compared with the surface forces, then the Hertz contact model could be used approximately. However, for microrobots and other micro/nanoscale systems, we need to avoid too high loads so that no damage or plastic deformation is caused on the surfaces because the contact stresses at a small contact area will be huge and could easily be higher than the yield strength of most materials. Therefore, we typically avoid high loads and apply loads comparable to the total surface force values. Then, the Hertz model is not appropriate to use any more, and we need to use the DMT, JKR, or MD model. To determine which of

Table 3.4

Micro/nanoscale elastic contact and adhesion (pull-off force) models for an atomically smooth spherical microrobot asperity in contact with an atomically smooth flat substrate

Model	Valid for	Pull-off Force	Description
DMT	$\lambda < 0.6$	$P_{po} = -2\pi RW_{12}$	Long-range surface forces act outside contact area
JKR	$\lambda > 5$	$P_{po} = -1.5\pi RW_{12}$	Short-range surface forces act inside contact area
MD	any λ	$P_{po} = -\alpha\pi RW_{12}$	Interface modeled as a crack ($1.5 \leq \alpha \leq 2$)

these three models is the most accurate one to use in a given microrobot case, the magnitute of the Tabor parameter given in Eq. [3.67] should be checked. For $\lambda < 0.6$, where the interface is not so adhesive (W_{12} is low), the materials are rigid (K is high), and the sphere radius R is small, the DMT model is an accurate model. In contrast, for $\lambda > 5$, where the interface is very sticky (W_{12} is high), the materials are relatively soft (K is low), and the sphere radius R is large, the JKR model is an accurate model. For any λ value, including the intermediate values between 0.6 and 5, the MD theory would give the most accurate deformation and adhesion models. An alternative intermediate adhesion model that is simpler to compute could be the Pietrement model [120, 123]. These cases are outlined in Table 3.4.

To determine λ, different material properties commonly seen in microrobotics are given in Table 3.5. For example, for a hydrophilic silicon microrobot with a spherical asperity in contact with a flat silicon surface, there would be water in between these two surfaces, which means $W_{12} \approx 2\gamma_{water}$. For $R =$ 1 mm, γ_{water} = 72 mN/m, a_0 = 0.2 nm, and K = 50 GPa, λ can be computed as 15.6. Then, for this case, the JKR theory would be accurate to model the contact deformation a, indentation δ, and adhesive force P_{po} using Eq. [3.61]. However, for the same parameters, if we change $R = 1$ μm by scaling down the sphere size from milli- to microscale, then $\lambda = 1.56$ and the MD model would be the most accurate model to use.

Due to the wide range of possible surface energies for glass given in Table 3.5, there is a large range of pull-off forces that can potentially exist with this surface. This wide range can be reduced when operating immersed in liquid.

3.5.1 Other contact geometries

The previous analytic models are valid only for a sphere-plane contact geometry, which could be sufficient for all microrobot adhesion modeling cases where

Table 3.5
Properties of materials commonly encountered in microrobotics studies

Material	Surface Energy/Tension γ (mJ m^{-2})	Elastic Modulus E (GPa)	Poisson Ratio ν
Glass	83-280 [124–127]	70	0.25
Polystyrene	33-40 [112, 128]	3.2	0.35
Silicon	46-72 [129]	160 [130]	0.17 [130]
Silicon dioxide	17.8 [129]	70 [131]	0.17
Gold	1,080 [132]	79	0.42
Nickel	2,450 [133]	200	0.31
SU-8	28-70 [102]	2.0	0.37
Polymers	~15-45 [134]	~0.3-3.4	~0.3-0.4
Elastomers	~15-100	~0.0001-0.1	~0.5
Water	72.3	-	-
Silicone oil	19.8-21.0 [135, 136]	-	-

the microrobot contact surface can be approximated by a sphere effectively and the robot is interacting with a planar or spherical surface (for sphere-sphere interaction, just replace R in the previous equations with $R_1R_2/(R_1+R_2)$, where R_1 and R_2 are the radii of each sphere, respectively). However, if the robot body shape is not possible to approximate with a sphere, then we need to derive adhesion models for more complex geometries.

From the literature, some useful adhesion models can be as follows. First, for a vertical cylinder contacting an atomically smooth flat surface, the adhesion can be computed using a circular flat punch model as [137, 138]

$$P_{po} = \sqrt{6\pi R_c^3 K W_{12}}\,, \tag{3.68}$$

where W_{12} is the effective work of adhesion between two surfaces, R_c is the cylinder radius, and K is the effective elastic modulus at the interface. Next, for a horizontal cylinder contacting a flat surface, the adhesion model can be given as [139]

$$P_{po} = 3.16 l_c (K W_{12}^2 R_c)^{1/3}, \tag{3.69}$$

where l_c is the cylinder length. Two horizontal cylinders in contact with an angle between them would have an elliptic contact deformation shape, and the adhesion is modeled approximately in [140]. Moreover, some preliminary elastic rough surface adhesion modeling can be found in [138].

3.5.2 Viscoelastic effects

All of these contact mechanics models assume no viscoelastic effects during contact or separation because such effects are complex to incorporate into the adhesion and deformation modelling [141]. For a very slow rate of retraction from a contacted surface, the previous elastic contact models are fully accurate. However, at higher retraction speeds, viscoelastic effects become important to model and take into account. At high speeds, viscoelastic effects could significantly increase the effective adhesion between a microrobot and surface.

Two main types of viscoelastic effects need to be taken into account. First, *bulk viscoelasticity* could increase the effective modulus of (stiffen) a given robot or surface bulk material during separation, which could effectively change adhesion if adhesion is a function of effective modulus of the materials in contact. For microscale bodies, except at very high speeds and for very lossy viscoelastic materials, bulk viscoelasticity could be generally negligible. Next, *interfacial viscoelasticity* is the most important dissipation mechanism at the micron scale, where high crack progation speeds at the contact interface could increase the effective interfacial work of adhesion, which would make the interface stickier. As a simple approximation of such interfacial viscoelastic effect, assuming negligible any bulk viscoelastic effect, we can define effective interfacial work of adhesion W as a function of crack propagation speed v as [142]:

$$W(v) = W_0 \left[1 + \left(\frac{v}{v_0} \right)^n \right], \tag{3.70}$$

where W_0 is the effective interfacial work of adhesion as v is approaching zero, n is a scaling parameter that is empirically fitted to the experimental data, and v_0 is the crack propagation speed at which W doubles to W_0. Such power law has been shown to hold for viscoelastic polymer materials at low crack speeds. To be able to use such a model, we need empirical measurements of W at different crack progation speeds by using force-distance measurements under an optical microscope where we can directly visualize the crack propagation at the interface. After fitting n to the data, we can replace the W_{12} values in all of these adhesion models with this $W(v)$ model in Eq. [3.70] to approximately include the interfacial viscoelastic effects. For example, in [142], fitted empirical data of the circular flat punch adhesion tests gave $n = 0.69$, $W_0 = 1.335$ J/m^2, and $v_0 = 3.37 \times 10^{-3}$ μm/s. Then, we can incorporate such values to

$W(v)$ and the adhesion model becomes:

$$P_{po}(v) = \sqrt{6\pi R_c^3 K W_0 \left[1 + \left(\frac{v}{v_0}\right)^n\right]}. \tag{3.71}$$

3.6 Friction and Wear

According to the Amonton's laws, friction is only load controlled at the macroscale (except special highly adhesive, non-rigid surfaces), and it is independent of the contact area. However, at the microscale, friction is also adhesion controlled (in addition to being load controlled) and thus contact area dependent. Therefore, we need to develop new models for friction at the microscale. A microrobot in contact with a given surface can slide, stick-slip, spin, roll, or do combinations of these motions at a given time, which are crucial to model and understand so that the designed microrobot can have a precise, stable, and energy-efficient motion control. In the following discussion, we will develop basic approximate friction models for such distinct contact motion modes.

3.6.1 Sliding friction

Assuming smooth and wearless surfaces, general sliding friction at the microscale can be modeled as

$$f = \tau_f A_f + \mu_f L, \tag{3.72}$$

where f is the friction force, τ_f is the interfacial shear strength, A_f is the real contact area, μ_f is the coefficient of friction of the interface, and L is the normal load. For low loads and highly adhesive interfaces, the Bowden-Tabor theory gives

$$f \approx \tau_f A_f, \tag{3.73}$$

which could easily be valid at the micro/nanoscale [143, 144]. Before the robot body starts sliding, f corresponds to a static friction force, with μ_f^s and τ_f^s being the static friction coefficient and interfacial shear strength, respectively. After sliding starts and reaches to its steady state, f corresponds to a kinetic friction force, with μ_f^k and τ_f^k being the kinetic friction coefficient and interfacial shear strength, respectively.

Depending on the size scale of the contact deformation, i.e., contact radius a, τ_f can be approximated by [145]

$$\tau_f(a) \approx \begin{cases} \frac{G}{43}, & a < 20\ nm, \\ G10^N(a/b)^M, & 20\ nm < a < 40\ \mu m, \\ \frac{G}{1290}, & a > 40\ \mu m, \end{cases} \tag{3.74}$$

where $M = tan^{-1}(G/43 - G/1290)/(8 \times 10^4 b - 28b)$, $N = 28b$, and $b = 0.5$ nm is the Burgers vector. Here, the bulk shear modulus of the interface G is defined as

$$G = \frac{2G_1G_2}{G_1 + G_2}, \tag{3.75}$$

where $G_i = E_i/2(1 + \nu_i)$, $i = 1, 2$.

The real contact area A_f is complex to calculate, and depends strongly on the surface roughness. Assuming the surfaces are smooth, we can calculate the real contact area as $A_f = \pi a^2$ by computing the contact radius a of the sphere-plane interface using the elastic contact mechanics models given in Section 3.5 and Table 3.4. After knowing A_f, the shear stress S during the sliding friction can be given as

$$S = \frac{f}{A_f} = \tau_f(a) + \mu_f P_i, \tag{3.76}$$

where P_i is the real local shear pressure. When two rough rigid objects contact, the real contact area would be less than the apparent contact area due to asperity contact. For cases when the normal contact force is dominated by adhesion, a common method to reduce friction is to reduce the contact area by adding small bumps or ridges on surfaces in contact to serve as contact points with reduced contact area [146].

3.6.2 Rolling friction

When a smooth spherical microrobot body rolls on a smooth flat surface as a locomotion mode, rolling friction resists its rotational motion. Such resistance is typically represented by the term M^r_{max}, maximum rolling resistance moment. When the interface is highly adhesive and compliant, i.e., the Tabor parameter λ is larger than 5, the Dominik and Tielens model [147] that used the JKR theory to predict the approximate analytic pressure distributions at the rolling interface can be used to approximate M^r_{max} as

$$M^r_{max} \approx 6\pi R \zeta_r W_{12}, \tag{3.77}$$

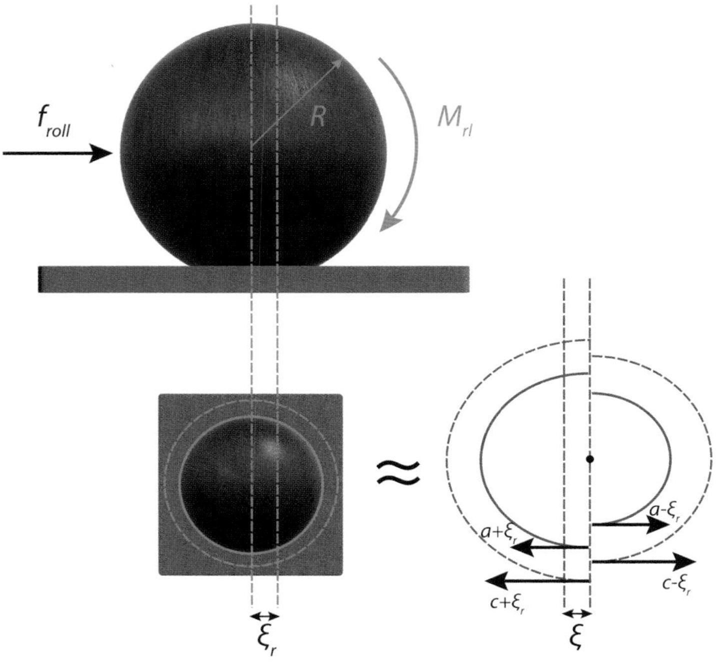

Figure 3.9
Side-view sketch of a sphere rolling on a plane due to an applied rolling moment M_r. Top view of the approximated shape of the rolling contact for a sphere on a plane.

where ζ_r is defined as the critical rolling distance and its value is in the range of $a_0 \leq \zeta_r \leq a$, where a_0 is the interatomic distance and a is the contact radius. For any λ value, as in the case of the MD theory, a general M^r_{max} solution needs to be developed. As driven by [145], by approximating the MD theory with a simpler to solve analytically but close solution (see Figure 3.9), M^r_{max} can be approximated as

$$M^r_{max} \approx \pi \zeta_r W_{12} \sigma_0 \frac{a^3 - c^3}{\sqrt{c^2 - a^2}}. \tag{3.78}$$

Here, c and $\sigma_0 = 1.03 W_{12}/a_0$ are given in the MD theory in Figure 3.8.

After knowing M^r_{max}, we need to apply enough external torque M_r that is larger than M^r_{max} to be able to roll the sphere on a given surface. Instead of an external direct rolling torque M_r, rolling moment could also be generated by an external lateral force f_{roll} along the center of the spherical robot body. In this case, f_{roll} needs to be larger than M^r_{max}/R to be able to roll it. Typically, such required f_{roll} is much smaller (e.g., two to three orders of magnitude) than the required sliding force f. Therefore, when the microactuation force magnitude is limited for moving a microrobot in contact with a surface using any given principle, such as in the case of magnetic actuation, it is much better to roll the robot on the surface because the required force could be several magnitudes order less than sliding.

3.6.3 Spinning friction

In the spinning motion, there would be a relative shear motion between the spherical body and the flat surface, as in the case of sliding motion. However, the particle rotates about its own center without a real lateral displacement in contrast to sliding motion. Therefore, the slip mechanism discussed earlier is valid, and the maximum spinning moment resistance M^s_{max} and the critical lateral force f_{spin} required to spin the particle at a given lateral x_0 distance from the center on the sphere surface can be written as:

$$M^s_{max} = \int_0^a 2r^2 \tau(a) dr = \frac{2}{3}\tau(a)a^3, \tag{3.79}$$

$$f_{spin} = \frac{2}{3x_0}\tau(a)a^3, \tag{3.80}$$

assuming a negligible vertical load. For non-negligible load cases, the load-controlled sliding friction term also needs to be integrated.

The required critical spinning friction f_{spin} is also typically much smaller than the sliding friction force f. Assuming $x_0 \approx R$ is selected, $f_{spin} \approx \frac{2a}{3R}f$. Because $a \ll R$ generally for microrobots and elastic contact models, $f_{spin} \ll f$. So, like rolling, it is typically much easier to move a microrobot on a surface by spinning rather than sliding.

3.6.4 Wear

Wear is a critical problem for sliding interfaces in micromechanisms, necessitating the use of special material pairings to achieve useful lifetimes in fast-moving interfaces [148]. Problems of wear, friction, and adhesion can be avoided through the use of flexure joints as opposed to rotary pin joints [149**?**] for miniature robot mechanisms.

3.7 Microfluidics

Many applications of microscale robots involve operation in fluids. Thus, the motion of microrobots by any method is subject to fluid forces. In this section, we cover the governing equations of fluid mechanics, focusing on simplifications for motion at the microscale. Of particular interest for microrobot motion, we introduce fluid drag relations for translating and rotating bodies.

Using the principles of conservation of momentum, a fluid flow is governed by pressure, viscous, and body (gravitational) forces. This balance of forces is assembled in the Navier-Stokes equation

$$-\nabla P_f + \mu \nabla^2 u + \rho g = \rho \frac{Du}{Dt}, \tag{3.81}$$

where P_f is the fluid pressure, μ is the dynamic fluid viscosity ($\mu = 10^{-3}$ kg/(m.s) for water and $\mu = 18.5 \times 10^{-6}$ kg/(m.s) for air at 20°C), u is the fluid velocity, ρ is the fluid density, and g is the acceleration due to gravity. Here, $\frac{Du}{Dt}$ refers to the material derivative of the fluid velocity. This governing equation assumes the liquid as a continuous, incompressible, Newtonian medium, and neglects effects such as Brownian motion. The continuum model of fluid mechanics is valid at all scales in the realm of microrobotics. Solutions to this equation, using appropriate fluid boundary conditions, give the fluid velocity vector field. In many applications, the gravitational (body) force term can be neglected, and we use a change of variables to analyze the relative magnitude of the remaining terms. Here, we use the non-dimensional variables $x^* = x/L_c$, $u^* = u/u_\infty$, and $P_f^* = P_f L_c / \mu u_\infty$, and divide by $\mu u_\infty / L_c$ to isolate terms on the left-hand side of the equation. Here, u_∞ is a characteristic velocity of the fluid (such as the free stream velocity) and L_c is a characteristic length, such as an object dimension. Thus, we arrive at the non-dimensional form

$$-\nabla P_f^* + \nabla^2 u^* = \left(\frac{\rho u_\infty L_c}{\mu} \right) \frac{Du}{Dt}. \tag{3.82}$$

The term in parentheses on the right-hand side of this equation is known as the Reynolds number $Re = \rho u_\infty L_c / \mu$ and is often interpreted as the ratio of fluid inertial to viscous forces. Thus, when Re becomes small, viscous forces dominate over inertial forces, resulting in creeping flow (Stokes flow), and can be thus described by the simpler equation

$$\nabla^2 u^* = \nabla P_f^*. \tag{3.83}$$

Fluid flow at the microscale, with small characteristic length L_c, is dominated by viscous forces as opposed to inertial forces. This flow would correspond to that seen at larger scales for low-density, slow, or high-viscosity flows.

Stokes flow has no dependence on time, and thus the solution for steady boundary conditions over all time only requires knowledge of the fluid state at a single time. In addition, the flow is time-reversible. Thus, reciprocal motions, where a motion and its opposite are repeated over time, will result in no net forces exerted on the fluid. This is known as the *Scallop theorem* [150] and is a notable difference when comparing microscale versus large-scale swimming methods.

The Stokes flow equation can be solved exactly by finding the Green's function (here called a stokeslet), numerically by a boundary element method, or by experimental characterization. The solutions to several interesting cases are now given. These fluid flow solutions are used in modeling microrobot motion through fluids as well as in studies of the fluid flow generated by moving microrobots.

3.7.1 Viscous drag

The fluid drag on any object at the microscale can be approximated by the drag on a sphere. As opposed to macroscale high-*Re* drag analysis, which includes multiple competing drag contributors such as viscous drag, form drag, and so on, Stokes flow drag is relatively simple because it results only from viscous forces. The drag force across a wide range of laminar flow can be found using the empirically derived Kahn-Richardson formulation for a sphere, F_{KR}, which is valid for a large range of small to moderate Reynolds numbers ($0 < Re < 10^5$) [151]:

$$Re = \frac{2\rho u R}{\mu}, \tag{3.84}$$

$$F_{KR} = \pi R^2 \rho u^2 \left(1.84Re^{-0.31} + 0.293Re^{0.06}\right)^{3.45}, \tag{3.85}$$

where Re is a function of the sphere's radius R, the fluid velocity u, the fluid density ρ, and the dynamic viscosity of the fluid μ.

For small Re, this force can be simplified using the viscous drag equation for a sphere at low Reynolds number [152], which provides results within 2.5% of the Kahn-Richardson model:

$$F_{drag} \approx 6\pi \mu R u. \tag{3.86}$$

Thus, the fluid drag force on a sphere in the low *Re* regime is proportional to the fluid viscosity, the sphere radius, and the fluid velocity. This simple model can also be applied to non-spherical shapes using a sphere of equivalent radius. Analytical solutions for such equivalent sphere radii for cases of ellipsoids, in addition to approximate correction factors for other geometries, are given in [153, 154].

3.7.2 Drag torque

Rotating bodies often appear in microrobotics. In a low *Re* environment, the drag torque on a rotating body can be solved exactly for an ellipsoid, and other shapes can be approximated by ellipsoids. The drag torque, assuming Stokes flow and an elliptical microrobot shape with major axis a and minor axis b, is given as [155]

$$\vec{T}_d = -\kappa_d V \mu \vec{\omega}, \tag{3.87}$$

where κ_d is the particle shape factor, given as

$$\kappa_d = \frac{1.6\left[3(a/b)^2 + 2\right]}{1 + \beta_1 - 0.5\beta_1 (b/a)^2}, \tag{3.88}$$

where

$$\beta_1 = \frac{1}{\epsilon^3}\left[\ln\left(\frac{1+\epsilon}{1-\epsilon}\right) - 2\epsilon\right] \tag{3.89}$$

and

$$\epsilon = \sqrt{1-(b/a)^2} \quad (a \geq b). \tag{3.90}$$

Here, $V = \frac{\pi a b^2}{6}$ is the volume of the microrobot and μ is the viscosity. Thus, the drag torque is proportional to the fluid viscosity and the rotation rate.

3.7.3 Wall effects

When operating in fluid near a solid boundary, the torque required to rotate or translate a micro-object increases. Liu et al. [156] studied a sphere of diameter D rotating about an axis perpendicular to a planar boundary in a low Reynolds number environment, giving a far-field approximation for the increase in torque from the unbounded fluid torque T_d to the torque in the presence of the wall T_w at a distance d from the wall. This ratio is given as

$$T_w/T_d = \left[1 - \frac{1}{64}\left(\frac{D}{d}\right)^3 - \frac{3}{2048}\left(\frac{D}{d}\right)^8\right]^{-1}, \tag{3.91}$$

with a value of 1.202 being the exact maximum torque ratio when in contact with the wall. For a sphere rotating about an axis parallel to the boundary, the far-field approximation gives

$$T_w/T_d \approx 1 + \frac{5}{128}\left(\frac{D}{d}\right)^3. \tag{3.92}$$

These far-field equations are accurate for values of $\frac{D}{d}$ greater than approximately 1.2.

For translation without rotation parallel to a nearby wall, the increase in drag force as a function of the wall proximity is approximated by [157, 158]

$$F_w/F_d = \left[1 - \frac{9}{16}\left(\frac{D}{d}\right) + \frac{1}{8}\left(\frac{D}{d}\right)^3 - \frac{45}{256}\left(\frac{D}{d}\right)^4 - \frac{1}{16}\left(\frac{D}{d}\right)^5\right]^{-1}, \tag{3.93}$$

which is valid for values of $\frac{D}{d} < 10$ (i.e., large distances from the wall). Approximations for near contact can be found in [158].

3.8 Measurement Techniques for Microscale Force Parameters

Geometric parameters in physical models such as sphere radius, 3D surface geometry and dimensions, and surface roughness can be easily characterized using optical microscopes, scanning electron microscopes (SEMs), atomic force microscopes (AFMs), and 3D optical scanners. Moreover, some physical parameters such as Young's modulus, Poisson's ratio, yield strength, liquid contact angle, Debye length, and surface charge density (zeta potential) parameters can be characterized using established measurement techniques. However, some microscale physical parameters such as interfacial work of adhesion W_{12}, real contact radius a (area, A_f), static and kinetic sliding friction coefficients μ_f and interfacial shear strengths τ_f, crack propagation speed v, critical rolling distance ζ_r, and maximum adhesive strength σ_0 of the interface are more challenging physical parameters to measure. We need to characterize adherence and friction forces and effective work of adhesion directly at the

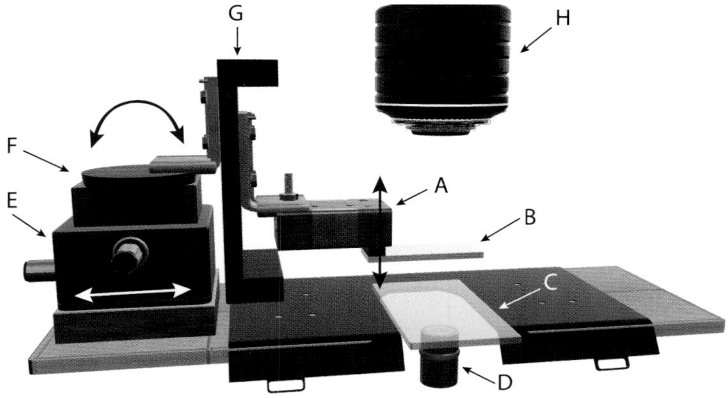

Figure 3.10

Schematic of an example custom force measurement setup to characterize adhesion and friction at the micron scale for mN scale interaction forces: A - load cell, B - (spherical, circular flat punch, etc.) indenter, C - flat surface, D - microscope objective, E - two-axis manual linear stage, F - two axis manual goniometer, G - motorized linear stage, H - light source.

microrobot and surface interface to know the exact force and energy values at the interface and validate and tune the proposed models because previous models are simplistic, neglect viscoelastic effects, and only approximate.

To measure these challenging physical parameters, a 2D force measurement system with precisely controlled 2D displacements and real-time or off-line visual feedback of the contact interface would be ideal. Force sensing needs to have resolutions in the range of a few nanoNewtons (nNs) to hundreds of microNewtons (μ.N.s) for single asperity and small area contacts while large area and multi-asperity contacts could require milliNewton (mN) scale force measurements.

For nN-μN resolution force measurements, an AFM with a spherical particle-attached nanoprobe could be used in air, vacuum, or liquids. The main challenge of typical AFM systems is not being able to image the real micro/-nanoscale contact area during the force measurement. Therefore, it can be mainly used for characterizing W_{12}, μ_f, and τ_f. ζ_r can be measured by an AFM system using a nanoprobe with a known shape and size tip to push and roll a spherical particle on a flat substrate (see [145]). However, as a special system, if an AFM system is integrated to an inverted optical microscope, as

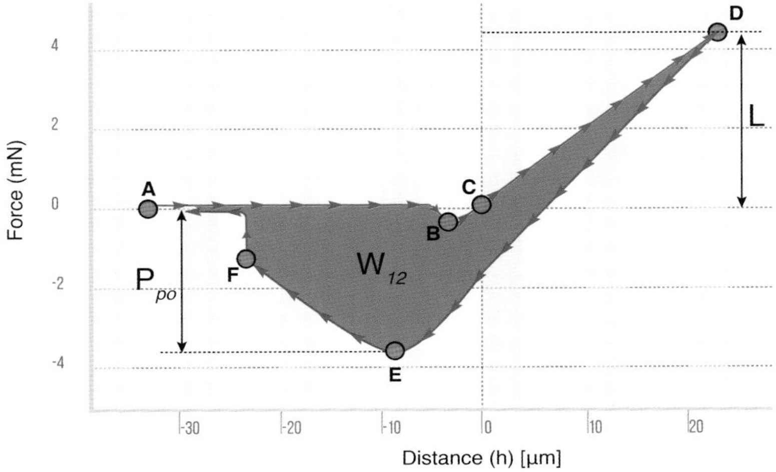

Figure 3.11

A possible force-distance curve between a smooth glass hemisphere loaded to and unloaded from a smooth, flat substrate. Two surfaces start approaching to contact from point A. Long-range attractive forces could attract the hemisphere with a maximum attractive force at point B. Two surfaces contact each other at point C when the distance reaches the interatomic distance a_0. The interface would deform elastically during compressive loading until reaching a predetermined load L at point D. An attractive (negative) maximum tensile force (i.e., pull-off force P_{po}) occurs at point E. Two surfaces separate from each other at point F.

in the case of many commercial AFM systems for biological imaging applications, microscale contact deformations can also be measured in real time on a transparent substrate.

For mN resolution force measurements, a custom 2D force measurement system can be built using two high-resolution load cells (for vertical and lateral force measurements) and two computer-controlled positioners integrated to an inverted optical microscope system for contact area imaging. A spherical, circular flat punch, or any other shape indenter can be moved on a transparent substrate with a controlled position and speed and contacted and retracted vertically for adhesion and contacted and laterally sheared for friction measurements. In Figure 3.10, an example custom force measurement system can be seen. Using such a setup, force-distance curve data can be generated as shown in Figure 3.11. Here, a hemispherical indenter with 1 mm diameter is loaded and unloaded from a flat array of circular, flat punch-shaped polyurethane elastomer fibers with 25 μm diameter and 100 μm length at a very low constant speed of 1 μm/s to minimize any viscoelastic effects. The spherical indenter contacts to the substrate at distance $a_0 \approx 0.17$ nm. After contact, the interface is compressed elastically depending on the effective modulus (compliance) of the interface K, which determines the slope of the force curve after contact. The indenter is stopped at a desired preload force at point B. Then the indenter is unloaded during points C-F with the same constant speed, where the tensile retraction force reaches a maximum tensile force at point E, which is defined as the adhesion of the interface P_{po}. At point F, two surfaces separate. Effective work of adhesion W_{12} and modulus K of the interface can be measured by fitting the measured data to the proper contact mechanics model. For a flat elastomer substrate instead of an array of fibers case, we can use contact mechanics models to fit such values. Assuming a case of $\lambda > 5$, where we can use the JKR theory, we can have the following relation between a and L:

$$\frac{a^{3/2}}{R} = \frac{1}{K}\frac{L}{a^{3/2}} + \left(\frac{6\pi W_{12}}{K}\right)^{1/2}. \tag{3.94}$$

If we plot $a^{3/2}/R$ as a function of $L/a^{3/2}$ from this equation, K could be computed from the slope of the loading portion of this plot [67]. After knowing K, it can be plugged in above equation and W_{12} can be found, where the Poisson's ratios and R of the indenter are assumed to be known.

3.9 Thermal Properties

Actuation in microscale mechanisms is often accomplished through thermal expansion of materials. The thermal strain ε_t induced from a change in temperature ΔT is

$$\varepsilon_t = \alpha \Delta T, \tag{3.95}$$

where α is the coefficient of thermal expansion. The value of α varies widely for different materials, meaning that dissimilar materials paired together will result in a bending along with expansion.

The heat lost through a surface due to conduction is proportional to the square of the surface area while the thermal energy it contains is proportional to the volume. Thus, a small object will conduct its heat quickly, requiring constant energy generation to maintain a high temperature. As a result, micromechanisms which are actuated by heating and cooling can be cycled quickly for high-speed operation. Thermal time scales with mesoscale objects can be on the order of minutes, whereas at the microscale these scales can be seconds or less.

3.10 Determinism versus Stochasticity

Stochastic events, e.g., those caused by thermal effects such as Brownian motion in fluid mechanics, have no significant impact on macroscopic systems, i.e., the actuation of the system easily overcomes stochastic events in the environment. But it is dominant, largely unexplored, nor constructively used for the control of microrobot perception-action cycles. In fluid dynamics, the Péclet number (Pe), defined in Eq. [2.10], is used to measure such effects. Pe $\gg 1$ would characterize macroscopic systems, which are largely deterministic in their behaviors. At small length scales, Pe < 1 means that transport is influenced by Brownian motion. The effect of everything being shaken around continuously requires new design principles for control at small scales, e.g., stochastic control of swarms of microrobots.

3.11 Homework

Design problem: How should you design a crawling microrobot on a flat substrate to have minimal sticktion to the substrate and have a controllable crawling locomotion? Discuss the details of how to choose the operation medium type and parameters and robot and substrate materials/coatings, surface texture, and geometry.

Force modeling problem: Insects and geckos have micro- or nanoscale foot-hairs to attach to a wide range of smooth and rough surfaces repeatedly and robustly. The following questions are about the adhesion of single and multiple polymer fibers with a spherical tip ending (such a spherical shape is assumed just for simplicity), inspired by these biological foot-hairs. Figure 3.12(a) shows a single polymer fiber with a tip radius of R on a flat and atomically smooth glass surface, and Figure 3.12(b) shows two fibers in contact with their tips, which is a possible configuration in many cases.

Use the default fiber and surface material parameters as below numbers unless they are mentioned to be different: E_{fiber} = 2 MPa, $\nu_{fiber} = 0.5$, $E_{substrate}$ = 70 GPa, $\nu_{substrate} = 0.3$, A_{fiber} = 80 zJ, $A_{substrate}$ = 200 zJ, $\gamma_{fiber} = 0.045$ J/m^2 (fiber surface energy, where the dispersion and polar components of the surface tension are $\gamma^d_{fiber} = 0.04$ J/m^2 and $\gamma^p_{fiber} = 0.005$ J/m^2, respectively), $\gamma_{substrate} = 0.2$ J/m^2 (substrate surface energy where the dispersion and polar components of the surface tension are $\gamma^d_{substrate} = 0.06$ J/m^2 and $\gamma^p_{substrate} = 0.14$

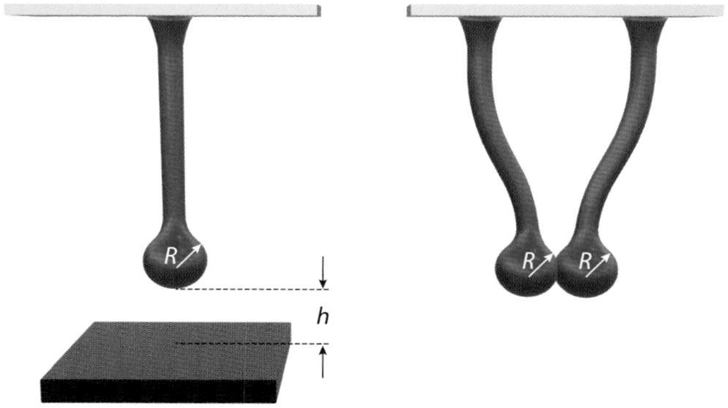

Figure 3.12

(a) Single polymer fiber with a spherical tip adheres to a flat glass substrate (left image), and (b) two polymer fiber spherical tips are in contact (right image).

J/m^2, respectively), and a_0 = 0.2 nm (interatomic distance when fiber tip and substrate or two fiber tips are in contact).

Until question (8), assume that the polymer fiber material is very hydrophobic (e.g., $\theta = 100°$). Therefore, these fibers can be assumed to use mainly van der Waals (vdW) forces as the dominant surface force to attach to surfaces until question (8). Using this assumption, answer the below questions:

1. For a fiber tip radius of R = 2 μm, calculate the single fiber tip adhesion due to vdW forces when the spherical fiber tip and an atomically smooth flat glass substrate contacts in ambient air. [Hint: For the fiber adhesion due to vdW force calculation, compute the vdW force when it reaches to its maximum value.]
2. Calculate the approximate fiber tip adhesion in air if the flat glass surface has a root-mean-square roughness of 20 nm (this is a similar roughness with an aluminum foil surface). Discuss the effect of roughness on adhesion from this result, and mention the possible physical reason for two rigid surfaces.
3. How would the fiber tip adhesion on glass change qualitatively if it is fully immersed in water? Next, calculate the exact fiber tip adhesion in this case (A_{water} = 37 zJ) and compare with the adhesion in (1) quantitatively.

4. Explain if there is any way to make the fiber tip and the substrate interaction due to vdW forces to be repulsive. If so, find and write down the relevant Hamaker constant condition to enable it.
5. For two fibers that are in contact with their tips during any surface adhesion test in air, as shown in Figure 3.12(b), what amount of spring force is required for one fiber to restore each fiber to its original vertical position?
6. In (5), if the two highly hydrophobic and compliant fibers in Figure 3.12(b) are fully immersed in water instead of being in air, what amount of spring force is required for one fiber to restore each fiber to its original vertical position? Here, derive an adhesion model for hydrophobic forces using the Derjaguin approximation, where the hydrophobic force energy is given by $W_{hydrophobic}(h) = -2\gamma\, e^{-h/\lambda_0}$ with $\lambda_0 = 1$ nm, and $\gamma = 0.072$ J/m^2 between two infinite planes. Compare this required force with the one in (5) and discuss.
7. Instead of computing the fiber tip adhesion from the van der Waals force models, it can also be computed by modeling the maximum force required to separate the fiber tip and the flat glass substrate. This separation force is also called pull-off force (F_{po}), which is determined by the micro/nanoscale contact mechanics model that you use, e.g., JKR, DMT, or MD model. Answer the following questions using the proper contact mechanics model for the fiber tip contact case in Figure 3.12(a):
   a. Determine which contact mechanics model (JKR, DMT, or MD) fits the best to the case in Figure 3.12(a) using the Tabor parameter. Then, compute the pull-off force, i.e., adhesion, using this model. [Hint: Neglect the water layer between the fiber tip and the glass substrate because the fiber material is hydrophobic.]
   b. For a zero vertical load, determine the friction between the fiber tip and a glass substrate by calculating the contact area using the contact mechanics model that you decided. Compute the interfacial shear strength between the fiber tip and the glass surface by assuming $\tau \approx G/30$, where $G = 2G_{fiber}G_{substrate}/(G_{fiber} + G_{substrate})$. For a friction coefficient of $\mu = 0.8$ between the fiber tip and glass substrate, plot and print the effect of preload on friction for preloads from zero to 1 μN.
8. If the polymer fiber material were hydrophilic with a contact angle of 20 degrees (i.e., $\theta_c = 20°$), then these fibers could use predominantly capillary forces to attach to surfaces in the air. For this case, calculate the fiber tip

adhesion on flat glass substrate due to capillary force for 90% humidity in the ambient environment for $R = 5$ μm and $\gamma = 72$ mJ/m^2. [Hint: To compute this adhesive force, use the h distance that gives the maximum capillary force value and take $d = 2r_k \cos\theta_c$.] Compare this adhesion value with the one due to vdW forces in question (1).

4 Microrobot Fabrication

Traditional robot fabrication relies on the use of bulk materials machined using mills, drills, etc.; rapid prototyping using 3D printers, laser cutters, precision assembly; and accommodates the incorporation of power, computation, and actuation on-board. These traditional techniques cannot be easily extended to the microscale, and so alternative microfabrication methods are used. Borrowing from the microchip and MEMS fabrication communities, microrobots are predominantly made using the methods of micromachining, including photolithography, material deposition, electroplating, micromolding, etc. Such micromachining techniques mainly consist of ultraviolet (UV) lithography, bulk micromachining, surface micromachining, LIGA (lithographie, galvanoformung, abformung) process, and deep reactive ion etching (DRIE) techniques. Additional processes using laser micromachining, two-photon lithography, electro-discharge machining (EDM), micromilling, etc. have also been explored.

Laser micromachining using commercially available or custom setups can cut almost any material with feature sizes down to tens of microns in 2D but suffer from speed as a serial process. It is typically used for 2D part geometries but can be used for some simple 3D cuts with low precision. For example, a magnetic microrobot can be cut from a bulk NdFeB sheet using a laser micromachining system (Quicklaze) as shown in Figure 4.1(a). EDM can likewise machine metallic parts down to tens of microns in size, as another serial process. EDM is typically limited to 2D part geometries. As a promising 3D fabrication process at the microscale, two-photon stereo lithography has been recently used to create high-resolution 3D shapes with features sizes of several microns out of UV-curable polymeric materials [159, 160], and can even be used for multi-material construction [161]. An example 3D-printed microrobot using a commercial two-photon lithography system (Nanoscribe) is shown in Figure 4.1(b), where a microhole at the center of top face of the cubic microrobot's body can trap a microbuble to pick and place a wide range of small parts in liquid media [162]. Moreover, such a 3D printer has been used to create 3D microrobot shapes which are functionalized using magnetic thin-film coatings [45] or embedded magnetic micro/nanoparticles [163]. Micromachining using small traditional cutters such as endmills operating at a high speed can also be used to create microscale features. These typically custom setups have been used to cut features tens of microns in size [164]. As a promising manufacture method for milliscale robots, the smart composite manufacturing (SCM) method can create layered mechanisms with integrated flexural joints with elements at the mm or cm scale [165]. These mechanisms are designed in

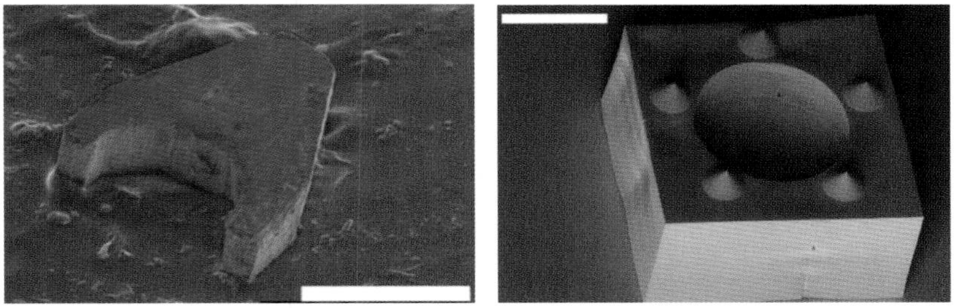

Figure 4.1

Scanning electron microscope (SEM) image of an example (a) magnetic microrobot fabricated by laser micromachining from a bulk NdFeB sheet to achieve strong magnetization properties, and (b) polymeric microrobot (coated with a magnetic Cobalt nanofilm) 3D-printed by a two-photon lithography system to trap a microbubble on the hole on the upper side of the cubic robot body for picking and placing small parts using capillary forces inside fluids. Scale bars are 200 μm.

2D, but are then folded to create complex 3D mechanisms [149]. However, this method will likely not scale down well to the sub-mm size due to the required assembly processes and reliance on complex layered designs.

Every microfabrication method that could be used to fabricate a microrobot has strengths and weaknesses, and needs to be selected depending on the given microrobotic application. Therefore, it is crucial to know all different possible techniques and choose the proper one in a given application. A summary of the features of different microfabrication methods relevant to microrobotics is given in Table 4.1. As can be seen, these commercially available techniques can fabricate different material types in 2D, 2.5D, or 3D with different resolution and aspect ratios. Here, 2.5D fabrication for a given 2D technique means that specific etching techniques can be used to create microstuctures more complex than 2D ones. In addition to the features given in this table, reproducability, yield, throughput, and cost are also important for a given fabrication technique towards especially industrial applications. Throughput, i.e., fabrication speed and volume, of serial fabrication techniques such as laser micromachining, two-photon lithography, micromilling, and EDM are low. These methods are more proper for rapid prototyping but not production for high-volume applications. In contrast, micromolding, EDM, two-photon lithography, and micromilling methods have yields much lower than 100% due to possible process uncertainties and time-dependent changes. Finally, most of these fabrication techniques require a basic clean room environment and expensive

Table 4.1
Features of different microfabrication techniques to fabricate microrobots or their components

Method	Dimension	Materials	Resolution (μm)	Aspect Ratio
UV lithography	2D/2.5D	Semiconductors, UV polymers	~0.5	Medium
Surface micromachining	2D/2.5D	Semiconductors, metals	~0.5	Low
Bulk micromachining	2D/2.5D	Semiconductors	~0.5	Low
DRIE process	2D	Semiconductors	~0.5	High
LIGA process	2D	Metals, polymers, ceramics	~0.5	High
Micromolding	2D/2.5D	Any (moldable material)	0.1-5	Medium
EDM	2D	Metals, semiconductors	5-10	High
Laser micromachining	2D	Any	5-10	Low
Two-photon lithography	3D	UV polymers	~0.1	High
Micromilling	3D	Any (removable material)	10-20	Medium

equipment, which could increase their cost for research and industrial applications significantly.

Fabricated microrobotic parts need to be assembled using self-assembly or precision robotic microassembly to create 3D functional systems. The techniques used depend on the functionality needed, with special materials required for magnetic actuation, and specific geometric components needed for all designs.

A major advantage of many microfabrication techniques is the parallel bulk nature. Commonly fabricated on a silicon wafer, hundreds or thousands of microrobotic parts are typically made on a wafer in a single process. These techniques typically only support 2D planar shapes, and the materials available for a particular process are limited. These techniques can be divided into wafer-level processes, and pattern transfer [166]. In this section, we also introduce other methods used in the fabrication of microrobots, including surface coatings, microassembly, and self-assembly, and we briefly cover the use of biocompatible materials for relevant applications.

4.1 Two-Photon Stereo Lithography

Microfabrication technology has reached to a state-of-the-art precision for almost any arbitrary 3D structure with the advent of the two-photon polymerization technique in 1997 [167]. Basically, the two-photon polymerization is the interaction of focused femtosecond laser radiation with a photosensitive material in a highly confined volume called voxel. This interaction leads to the polymerization (or curing) of the material within the voxel in a high temporal and spatial resolution, with the current minimum feature size down to 80 nm. This technology has become accessible by commercial turnkey systems, such as Nanoscribe and Workshop of Photonics, for microrobotics and many other researchers in different fields.

Microfabrication with the conventional photolithography relies on single-photon, i.e., UV-curable, 2D patterns of the photosensitive material limited only to axially uniform cross-sections because light passes vertically through the entire sample. To render three-dimensionality, layer-by-layer growth is the pursued strategy, which results in the loss of fidelity in the vertical dimension during the iterative procedures. The two-photon lithography, in contrast, benefits from the transparency of the photosensitive material in the near-infrared radiation while being highly absorptive in the UV range. This is successfully exploited for the initiation of polymerization precisely at the focused near-infrared femtosecond laser pulses. Therefore, the photosensitive material is polymerized along a trajectory assigned for the sample stage moving with respect to the laser focus, in all three dimensions, and thereby enabling axially complex 3D structuring by *direct writing* in the volume of photosensitive material.

In the step following the polymerization, the non-reacted polymer is removed by developing the sample in a proper solvent, leaving the fabricated structure on the supportive substrate. Consequently, the two-photon lithography successfully accomplishes the 3D printing of complex computer-aided designs in a highly practical fabrication scheme with high precision and sub-micron resolution, thus enabling an unmatched tool for realizing almost any desired microstructure. Nevertheless, comparatively low writing speed, small writing work space, and impractical size scaling are the major setbacks that need improvements for the currently available two-photon lithography-based microprinter systems.

Typically, the effective maximum writing speed depends on the available laser power and photosensitivity of the material, so no conventional recipe includes an applied laser dose and maximum writing speed, which fit to all materials. As a result, each material needs to be optimized for desired structural quality and writing speed. In the hardware side, however, the writing speed and accuracy are controlled by the piezo actuators in xy- plane and z-dimension, where the sample is moved relative to the fixed laser beam. Despite the high spatial resolution achieved by the piezo stage (i.e., on the order of a nm), the maximum achievable writing speed offered by the piezo system, typically between 30 and 150 μm/s, is well below the desired practicability for rapid production of bulky microstructures. To improve the writing speed, an alternative approach is to laterally scan the *xy*- plane by galvanometric mirrors while controlling the *z*-dimension by the piezo stage. This moving-beam fixed-sample approach allows for several orders of magnitude higher fabrication speeds (up to m/s) by employing a layer-by-layer building process [168]. Further, galvanometric scanning is subject to aberrations and vignetting, which dramatically reduce the accuracy for writing the edges of large structures within the field of view of the objective.

The two-photon lithography has enabled new functional photonics devices, including aperiodic photonic structures and mechanical metamaterials, 3D complex tissue scaffolds, and microfluidic devices and filters. In microrobotics, it has been used to fabricate new microswimmers to transport and deliver a specific cargo [45, 169].

A microswimmer should be optimally designed to experience minimum drag with the surrounding body fluid so as to gain maximum propulsion efficiency. Further, the microrobot should contain pre-programmed porosity for storing therapeutics with different sizes, i.e., ranging from a few nm to tens of microns. Such structural considerations can be potentially addressed by the two-photon lithography toward optimally designed, spatially heterogeneous microrobots for the most efficient functionality. As an early example of such a 3D-printed microrobot, a magnetic helical microswimmer, inspired by bacterial flagellar movement, was rotated and propelled using remote magnetic fields [45]. The helical design enabled cork-screw shape movement under a rotating magnetic field, resulting in a translational movement along the direction of the screw axis.

Responsive or smart materials may also presents a new opportunity for microrobot fabrication. For example, a liquid-crystalline elastomer fabricated by two-photon absorption direct laser writing with sub-micrometer resolution maintained the desired molecular orientation, which was then exploited to develop a microwalker by optical actuation [170]. From the applications side, novel materials having two-photon polymerization compatibility need to be developed for achieving advanced functionalities. In particular, development of new biomedical tools requires new biodegradable, biocompatible, and, if possible, bioactive materials. Direct 3D writing of conductive, metallic, magnetic, and elastomeric materials open the way to a wealth of possible directions.

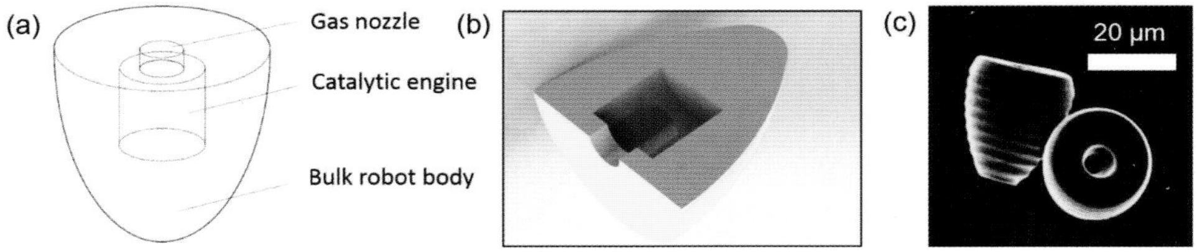

Figure 4.2

An example 3D microrobot design fabricated by two-photon lithography. (a) Each component can be selectively addressed by the two-photon laser pulse, thereby allowing for selective functionalization via light chemistry. The catalytic engine wall is selectively patterned with platinum nanoparticle groups that produce gas microbubbles inside the engine. The microbubbles leave the robot through the nozzle, inducing a thrust that propels the microrobot forward. (b) Section view of the catalytic microrobot shows its components in a 3D CAD drawing. (c) SEM micrograph of an example microrobot fabricated from a biocompatible polymer, polyethylene glycol diacrylate.

Along with the 3D structure, tailoring the local chemical properties by 3D patterns and gradients provides a further degree of control to insert complex functionalities into microrobots [171]. The chemical heterogeneity could be achieved in two ways. First, bulk parts of a microrobot are made of materials with dissimilar physical or chemical properties, so each part serves as a separate component to carry out its prescribed tasks. For example, one component may contain embedded magnetic nanoparticles inside a densely crosslinked network for remote steering while another component may contain drugs, so that the cargo will be released only at a tumor site as a response to the local pH change. Such complexity in a microrobotic design scheme has not yet been addressed. Second, patterning microrobots by grafting certain (bio)chemical moieties may enable extended surface and bulk functionalities in well-defined, confined sites in the microrobot.

For example, selective 3D surface patterning of catalytic platinum nanoparticles onto the inner cavity of a 3D microswimmer could create a complex propulsion system bubble propulsion, as shown in Figure 4.2. To realize these strategies, the two-photon phenomenon is a useful tool because it effectively localizes the reactive light dose in small voxels for grafting molecules or bonding large bulks at the interfaces. Nevertheless, bonding dissimilar materials requires compatible light-sensitive groups to be present on each component because the laser light provides the energy for reaction to initiate, but the bonding is facilitated between the compatible molecular species. This problem could be addressed by either exploiting the unreacted bonds during the

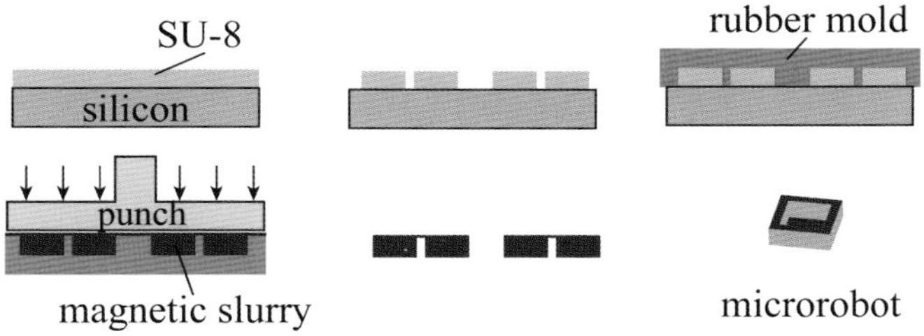

Figure 4.3

Replica molding process based on photolithography used to fabricate a large number of magnetic microrobots. From top left to bottom right, process steps are: deposit an SU-8 photoresist on a silicon wafer; pattern the SU-8 layer using UV lithography; replicate the SU-8 pattern's negative using a silicone rubber mold; mold the rubber mold with a liquid polymer mixed with magnetic microparticles and use a punch to remove the excess polymer mix from the mold; after curing the polymer, demold the microrobots. This process allows for the creation of arbitrary 2D-shaped polymer composite microrobots from micron to mm scale.

initial polymerization of the microrobot, or the photosensitive material could be prepared, such that consecutive steps enable light-sensitive new bonding sites.

4.2 Wafer-Level Processes

Wafer-level processes include cleaning and materials deposition. The first method of deposition is physical vapor deposition (PVD), which can be evaporation or sputtering. Evaporation can be used to deposit metals, and is commonly used for seed layers or electrodes. Sputtering uses an inert gas plasma to knock atoms from a target surface, which then reach the substrate to be coated.

Electroplating is used to deposit ions from solution onto a surface. Gold, copper, chromium, nickel, and iron-nickel magnetic alloys are commonly deposited with electroplating, and this method is capable of forming thicker layers than PVD, sputtering, or chemical vapor deposition.

Spin casting is a simple mechanical method of spreading a drop of liquid onto a wafer by spinning the wafer. Centrifugal forces balance with the solution surface tension to form a uniform thickness dependent on the spinning rate. This is the typical method to apply photoresist, which is later used in photolithography, and can form microrobotic structures or act as a masking layer for further processes.

4.3 Pattern Transfer

Microrobot designs can be transferred to the materials used by pattern transfer. In the commonly used optical transfer, light is selectively shown through a patterned photomask onto the substrate material. This substrate material is then selectively altered by the patterned light. One common optically sensitive material is photoresist, which forms a cross-link bond in the presence or absence of light. Thus, regions exposed by the mask become etch resistant or prone in subsequent chemical etching steps, leaving behind only an extruded structure corresponding to the mask shape.

Soft lithography is another method of pattern transfer which uses polymer molds to transfer patterns [172]. Poly(dimethylsiloxane) (PDMS) or other rubbers are often used as the mold, and the master pattern is often created using lithography and etching from photoresist or silicon. Final parts can be created from polymers using the mold (shown in Figure 4.3) or the mold can be used for transfer printing of thin films. For creating more advanced magnetic microrobots with an integrated compliant gripper that could open and close by remote magnetic field control, a molded elastic robot body with embedded magnetic microparticles can be magnetized, as shown in Figure 4.4 [26]. Soft lithography allows for reusable molds, and offers a simple, rapid method.

If the molded robot body is composed of an elastomeric material with embedded magnetic microparticles, as in Figure 4.3, then such a *soft* magnetic microrobot can be magnetized while the robot body is deformed in various shapes to create more complex soft microrobots with not a uniform but a spatially different magnetization profile. For example, as shown in Figure 4.5, a rectangular elastomer microrobot body with embedded NdFeBr microparticles is rolled around a cylinder and then magnetized in a given direction, which could enable a soft swimming robot that can create an undulation on the robot body as a traveling wave to propel on or under water efficiently [25]. Extending such a method to different static and dynamic deformations using a magnetoelastic sheet with more complex magnetization profiles has become available recently for complex programmable soft robots at the small scale [173].

Another fabrication method involves the inclusion of flexible elastomer elements in silicon features [174]. This allows for the creation of flexure hinges and elastic energy storage elements in a traditional MEMS process. This or similar processes could greatly increase the design freedom for microscale mechanisms in microrobot applications.

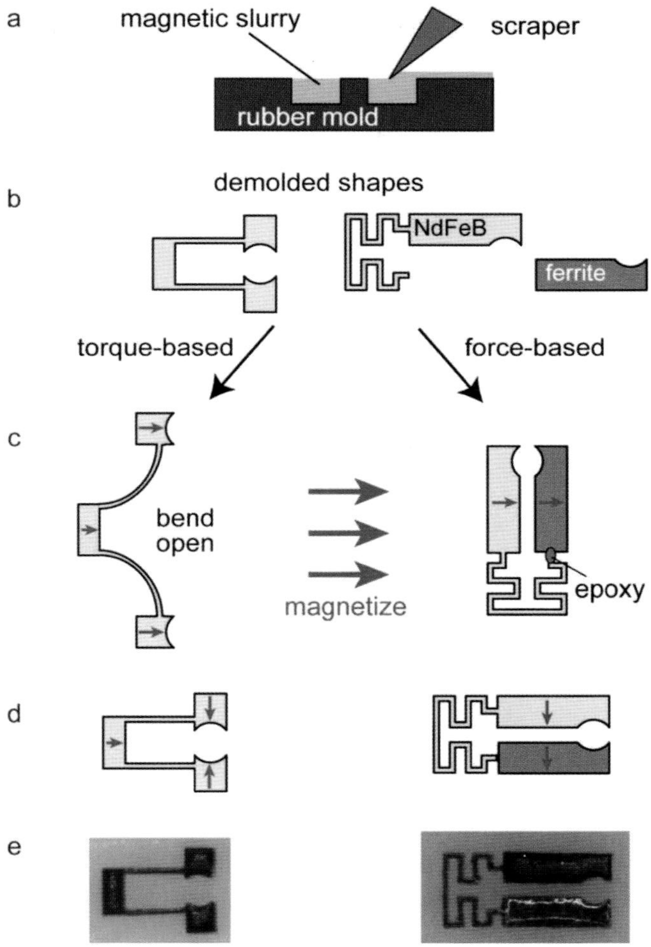

Figure 4.4

Fabrication and magnetization process for two magnetic microrobots with an integrated compliant flexural gripper [26]. Copyright © 2014 by John Wiley Sons, Inc. Reprinted by permission of John Wiley & Sons, Inc. (a) A magnetic slurry consisting of magnetic microparticles and polymer binding matrix is poured into the negative mold. (b) Microgripper shapes are pulled from the mold using tweezers. (c) Torque-based designs are spread open prior to magnetization, to allow each gripper tip to be magnetized in an opposite direction. The bend direction shown results in a normally closed gripper. Force-based micrograppers are molded from two magnetic materials, in two separate molding batches. The pieces are fixed together using UV-curable epoxy with a rubber mold as a fixture to hold the parts precisely. These force-based gripper tips are magnetized in one common direction. (d) After relaxation, the grippers are shown in their final magnetic configurations. (e) Fabricated designs are shown in the relaxed state after magnetization and assembly.

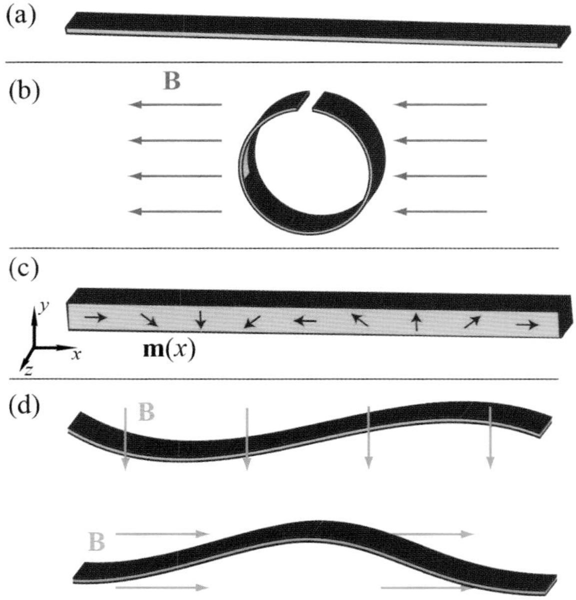

Figure 4.5

Fabrication, magnetization, and actuation processes of a swimming-sheet magnetic soft microrobot. Reprinted from [25] with the permission of AIP Publishing. (a) A flat sheet fabricated from permanent magnetic microparticles and Ecoflex silicone rubber is (b) bent into a circle and subject to a 1.0 T uniform magnetic field. (c) When the field is removed and the elastic robot body is straightened, it is left with a magnetization that varies along its length, which (d) causes it to be deformed when subject to a weak external field. Rotating the external field continuously in time causes the sheet deformations to travel down its length, providing a propulsive force in fluid.

Table 4.2

List of potential probes, targets, and binding methods for the chemical functionalization of a microrobot sensor surface

Target Molecule	Probe Chemical	Binding Method	Ref.
CO_2	Acrylamide + isooctylacrylate	Polymer matrix	[175]
NH_3	Poy(arcylic acid-co-isooctylacrylate)	Polymer matrix	[175]
H^+ (pH)	Acrylic acid + iso-octyl acrylate	Polymer matrix	[176]
Glucose	PVA and a co-polymer made of DMAA, BMA, DMAPAA, etc.	Polymer matrix	[175]
H_2O (Humidity)	TiO_2 or Al_2O_3	Solgel deposition	[175]
B. anthracis spores	anti-*B. anthracis* phage	Gold	[177]
fPSA	Rabbit anti-human PSA	Evaporated gold + DTSSP	[178]
Streptavidin	Biotin	N/A	[179]
Monoclonal antibody to (2,4-D)	2,4-dichlorophenoxyacetic acid	Evaporated gold + cysteamine + BSA	[179]
Myoglobin	Anti-myoglobin monoclonal antibody	sulfo-LC-SPDP	[179]
LDL cholesterol	Heparin	Evaporated gold + 2-aminoethanethiol-HCl	[180]
Immoglobulin G (IgG) and BSA	Gold	Evaporation	[180]

4.4 Surface Functionalization

Magnetic or electric functionalization of microrobot parts can be achieved through material choice and coatings. For magnetic actuation, a magnetic material such as nickel, iron, cobalt, or their alloys can be included as a bulk material, or as particles mixed into a polymer binder. Electrodeposited nickel or iron-nickel (Permalloy) can be patterned and formed to thicknesses up to several hundred microns [71]. Ground particles of iron, iron oxide, or rare earth magnet materials can be included directly with photoresists [39] or in molded plastic parts [13].

For electrical functionalization, microrobot features can be made electrically insulating using a polymer construction, or conductive using metallic construction or coating. Other functionalizations can be applied for passive remote sensing by chemical patterning, as shown for an optically excited and read oxygen sensor coating [181]. For chemical, biological, gas, or other sensing applications in a liquid or gaseous environment, during or after the robot's sensor fabrication process, the sensor element's surface needs to be functionalized by a specific chemical probe, which could binds specifically with a target molecule. The attachment procedure may require an intermediate binding layer, such as a polymer matrix or a deposited metal layer, as probe molecules may not readily adhere to the sensor material. A list of probe molecules, target molecules, and binding materials used in the microcantilevers literature is listed in Table 4.2. Probes and targets for such microcantilevers can potentially also be used on mobile microrobots.

4.5 Precision Microassembly

While it is simplest to fabricate microrobots in their final form using bulk manufacturing techniques, it is also possible to assemble parts, especially to achieve out-of-plane 3D features. Manual methods such as tweezers can be used to assemble parts down to roughly 100 μm in size, but precision assembly is only accomplished by robotic micromanipulation systems. In [71], multiple planar electroplated nickel parts of several hundred μm in size are assembled using such a system. These microassembly methods are serial processes which may not be compatible with bulk fabrication.

4.6 Self-Assembly

An alternative to precision robotic assembly is so-called self-assembly, where microscale interaction forces between parts cause parallel assembly. Such behaviors have been shown for specially designed parts using capillary, magnetic, and electric forces at the microscale [182]. Compared with microassembly techniques, generally self-assembly can be done smaller, faster in parallel, and in a self-correcting manner.

Self-assembly can be particularly useful in microrobotics to overcome the common limitation of 2D fabrication methods for microscale components. As one example of this, self-folding patterned 2D sheets have been shown to create complex 3D shapes, as shown in Figure 4.6 [183], and shapes with advanced electrical characteristics [184].

Such capabilities could be used to create functional 3D microrobot features for locomotion, sensing or form tools for manipulation of microscale parts at size scales down to tens of micrometers.

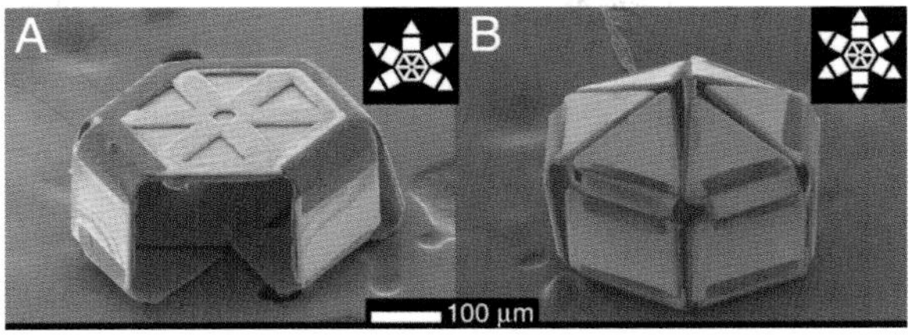

Figure 4.6

Scanning electron microscope images of self-folded untethered microgrippers from 2D patterns using directed self-assembly. Reprinted from [51].

4.7 Biocompatibility and Biodegradability

For biological applications in fluidic chips or medical applications inside the fluid cavities of the human body, biocompatibility of microrobots is a major concern. Many materials commonly used in microrobot fabrication are not biocompatible, including most magnetic materials. Most microrobot studies do not address biocompatibility, with one exception being [45], where mouse myoblasts are shown to grow on microscale IP-L and SU-8 photoresist microhelices. The use of surface coatings could render microrobots made from other materials biocompatible, although further study is required. For example, parylene or polypyrrole, common biocompatible polymers, could be coated over the other functional materials [185–187]. The use of such coatings in microrobotics is just beginning to be investigated, and warrants further study.

For some potential biomedical applications inside the human body in the future, the microrobots might not be collected or discharged easily. In these cases, microrobots could be made of biodegradable materials, in addition to being biocompatible, as one potential solution [55]. Therefore, microrobot materials need to be selected from biodegradable polymers, hydrogels, and other synthetic or biomaterials so they can degrade themselves in the given physiological environment without giving any side effects. Moreover, the immune system could attack such microrobot materials depending on the physiological environment, and therefore immunotoxicity type of issues must be prevented with the proper surface functionalization.

4.8 Neutral Buoyancy

For microrobots moving in 3D, the ability to levitate is complicated by the need for microrobot weight compensation. Thus, it could be desirable to achieve neutral buoyancy in a microrobot. In [16], magnetically actuated untethered CoNi microrobots are studied for complex motion in 3D liquid environments. For weight compensation, vertical forces must always be applied in addition to control forces, negatively influencing motion control. In the vertical orientation of the applied magnetic field, a 300-μm microrobot of buoyant weight 0.95 μN in water experiences a maximum upward magnetic driving force of 2.2 μN. Therefore, the weight of the microrobot is around 25% of the maximum force that the magnetic system can deliver in that direction. The absence of weight compensation could allow the application of more equal forces in every direction and hence better movement. Neutral buoyancy in magnetic microrobots is especially difficult because magnetic materials are dense. The approximate densities of relevant materials are listed in Table 4.3.

Low-density hydrogels can be used to lighten a microrobot [191], and 3D magnetic microstructures can be created featuring low-density hollow glass microcapsules for buoyancy [163]. These are mixed with photocurable polymer and formed using a laser into intricate 3D shapes such as helices. In medicine, blood, cerebrospinal fluid, and the urinary tract are all regions of high focus given the potential benefits of microrobot technology. Each of these fluid environments has a density value close to water (1,000 kg/m^3), and so this is often chosen as a microrobot target density. A similar fabrication method can take advantage of large trapped air cavities for buoyancy [192]. This design is fabricated by photolithography to include a large cavity, which is capped manually using polymer caps. This design can be tuned to the correct density, and consists of only magnetic material and binding polymer.

To make the neutrally buoyant robot designs easier in water, Percoll medium could be mixed with water to increase its density. For example, 140 μl of Percoll (Sigma-Aldrich, St. Louis, MO) mixed with water could increase the water's density to 1,090 kg/m^3 [23]. Such denser medium is compatible for biological applications.

Table 4.3

Densities of common materials in microrobotics studies at 25°C for synthetic fluids and body temperature for biological fluids

Material	**Density** ρ (kg/m^3)
Water	997 at 25°C
Silicone oil	971 at 25°C
Glycerol	1,260
Human blood	1,043-1,057 [188]
Human urine	1,000-1,030
Human vitreous humor	1,005-1,009 [189]
Human cerebral spinal fluid	1,003-1,005 [190]
Air	1,184
NdFeB	7,610
Iron	7,870
Nickel	8,900
Cobalt	8,900
Gold	19,320
Titanium	4,430
Platinum	21,450
Silver	10,490
PDMS	965
Polystyrene	1,050
PEG 400	1,128
Polyurethane	~1,200

4.9 Homework

1. List the possible methods from the literature to fabricate spherical Janus particles in large numbers, where half of the particle is coated with a different material, e.g., metal. Which self-propulsion methods can be used to create microswimmers using such Janus particles inside proper chemicals?
2. Which microrobot fabrication techniques have been used to create magnetic microswimmers with a helical tail or body with less than 1 mm overall sizes? What is the smallest size synthetic helical swimmer fabricated so far?
3. Which hard magnetic microparticle materials are available in the literature to fabricate magnetic microrobots with high magnetization properties? What is the smallest size of such micoparticles while preserving their hard

magnetic properties? Is there any biodegradable micro/nanomagnetic particle material with no coating? Are there any chemical or biological materials that demonstrate magnetic properties?

4. Find and list the microrobots in the literature that were fabricated by laser micromachining, self-assembly, and two-photon stereo lithography, respectively. Also list the proposed potential applications of each proposed microrobot.

5 Sensors for Microrobots

Mobile microrobots could have on-board or off-board (remote) sensors for a given application. A conceptual drawing of a microrobot capable of on-board sensing and interacting with entities within its environment is shown in Figure 5.1. Typical design parameters for a microsensor are:

- *Resolution*: Defined as the smallest change that can be detected by the sensor. It is typically determined by the electrical noise in the electronic output. Sensor resolution should be at least 10 times smaller than the parameter to be measured as a general rule of thumb. For example, if we like to detect a 1 μN contact force with a surface that is slowly changing, then the force sensor must be able to detect at least 0.1 μN at low frequencies. Resolution is a function of the measurement bandwidth/frequency; therefore, to detect a parameter change at a given frequency, the sensor needs to have the sufficient resolution at that frequency.
- *Sensitivity*: Defined as the change in output of the sensor per unit change in the parameter being measured. The factor may be constant over the range of the sensor (linear), or it may vary (non-linear).
- *Range*: Every sensor is designed to work over a specific range. When the resolution of the sensor is much higher, its range is typically much lower.
- *Bandwidth*: Bandwidth, i.e., frequency response, is an indication of the sensor's ability to respond to changes in the measured parameter. For example, the sensor's bandwidth should be high enough to detect fast measured parameter changes.
- *Noise*: The electrical noise in a sensor's output is the primary factor limiting its smallest possible measurement. Therefore, it should be minimized. A quantitative metric for noise is the signal-to-noise ratio (SNR) of the sensor, which should be maximized.
- *Non-linearity*: Sensor response could have non-linearities, including hysteresis. Such non-linearity could complicate the sensor calibration and feedback control using such sensors.
- *Power consumption*: Power consumption of a given sensor (and all other on-board components) should be minimized because small mobile robots have a limited power source.
- *Compatibility to the operation environment*: Depending on the application, mobile microrobots need to operate in air, water, body, or other physilogical fluids, or vacuum at low or high temperatures, pH, flows, etc. Therefore,

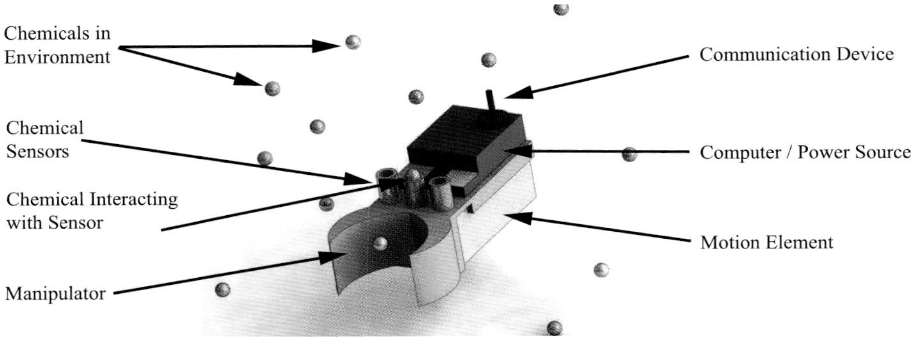

Figure 5.1
Conceptual drawing of a microrobot capable of sensing and interacting with entities within its environment. In this example, the microrobot could be capable of detecting chemicals through the use of chemical sensors and could be able to transmit these data wirelessly. A microrobot may also have a manipulator to actively control objects in its environment.

their sensors with given material selections need to be compatible to such environmental conditions.

- *Size*: The overall size of the sensor should be much smaller than the overall size requirement of the untethered microrobot for a given application. Therefore, the sensor and its processing electronics should have minimal size, and they should be integrated to the robot in a compact manner.
- *Weight*: For specific applications such as flying or levitated microrobots, the weight of the sensor needs to be minimized.

Small-scale mobile robots need many different types of sensors, such as image, displacement/strain, force/pressure, acceleration, angular rate, biological, chemical, gas, flow, temperature, and humidity sensors for a given application. This chapter covers the possible image sensors and transduction methods for possible various sensor designs.

5.1 Miniature Cameras

Vision is the most important non-contact (far-field) sensing modality for animals and robots to operate in complex environments. Therefore, imaging sensors are indispensible if available for a miniature robot at a given size scale. Like the mammalian eye, current high-resolution miniature cameras generally rely on a single lens, focusing the images onto a photosensor (retina). Such a lens could actively focus images, as in the case of current mobile phones using electrowetting liquids [193], similar to ciliary muscles altering the tension on the biological lens, changing its curvature and its focal length. The imaging bandwidth of a conventional camera is 20-50 frames per second (fps), limited by the data transfer rate and image-capturing hardware. The human eye has a resolution of 120 megapixels (11,000 × 11,000 pixels) while the current miniature cameras have typically low resolution (62,500-410,000 pixels). In contrast, small insects have compound eyes composed of hundreds or thousands of lenslets. They have a small size and low resolution (up to 100 × 100 lenslets) and are optimized for the detection of shape and motion at high speeds (e.g., 120 fps). Apposition and superposition types of lenslets collect and filter light to create non-focused, blurred images. Several groups have prototyped synthetic compound eyes using microfabrication techniques towards use in miniature robots and devices [194].

Current image sensors have two main types: charged coupled device (CCD) and complementary metal oxide semiconductor (CMOS). CCD sensors consist of an array of photodiodes, detecting the electrical charge accumulated proportional to the light intensity projected on them. Such an electrical charge on each photodiode is sequentially detected and sent to a separate digital signal processor at a given frame rate. In contrast, microfabricated CMOS sensors have an array of photodiodes that have on-chip, integrated electronics and processing, which make them low power consuming, miniature, light, and cost-effective. Comparing these two image sensors for miniature robot applications, CMOS sensors have slightly lower image resolution, less color reproducibility, darker brightness dynamic range (i.e., CMOS sensors require double power as CCD ones do to cover its darkness), shallower depth of field, smaller size, and lower total power consumption than CCD sensors. Such tiny CMOS and CCD sensors-based cameras have been commonly used in FDA-approved milliscale wireless endoscopic capsules (Given Imaging and Olympus), with 10 mm diameter inside the gastrointestinal tract. The current tiniest commercial camera (Naneye cameras, Awaiba) uses CMOS image sensors and has 1 mm^3 size, 250 × 250 pixels resolution at 42-55 fps, and 4.2 mW power consumption. For medical applications inside the human body, single or multiple light sources such as light emitting diodes (LEDs) are required on the robot to have images, which would increase the size and power consumption of the image sensing system.

5.2 Microscale Sensing Principles

Microscale sensing principles could be capacitive, piezoresistive, optical, piezoelectric, magnetic, thermoelectric, thermomagnetic, etc. These transduction methods, except the piezoresistive one, also can be used for actuation. The most common ones for microrobotics are capacitive, piezoresistive, piezoelectric, and optical. Taking a silicon microcantilever probe with a nanoscale sharp tip at the end, used commonly in atomic force microscopy (AFM), as an example microsensor that could be integrated to a microrobot, capacitive, piezoresistive, and optical principles have been mainly used to detect the static or dynamic deflection of the cantilever probe. In this example case, optical, capacitive, and piezoresistive deflection sensing systems have around 0.1, 0.01, and 1.0 nm resolution as a basic resolution comparison case. As another example, a microfabricated cantilever beam can be modified to measure temperature in an environment by constructing a bilayer cantilever made of two materials with different coefficients of thermal expansion, which induces mechanical deflection due to temperature change that can be detected using capacitive, piezoresistive, and optical principles [179]. Piezoelectric principle is not used to detect static deflections typically but can detect oscillatory deflections precisely. Such a principle will be discussed in Section 6.1 when it is described as an actuation method.

Recent developments have also enabled highly stretchable soft strain and other sensors at the small scale using resistive or capacitive sensing principles of an elastomer with embedded different nanomaterials, such as nanowires, carbon nanotubes, and nanoparticles [195, 196]. Such soft sensors would be crucial to be integrated into future soft milli- and microrobots.

5.2.1 Capacitive sensing

Capacitance between two parallel conductive plates would change when one of the plates moves vertically or horizontally. For example, putting one of these electrodes to the robot body and the other one to the surface that it moves on could enable the robot distance or contact sensing with the surface. The capacitance between two plates C when they have a distance h in between is

$$C = \frac{\varepsilon \varepsilon_0 A}{h}. \tag{5.1}$$

Here, A is the plate surface area and ε is the relative dielectric constant of the material between the plates and ε_0 is the permittivity of free space (8.85×10^{-12} F/m). When one of the plates moves vertically with a distance Δh, the capacitance change ΔC would be

$$\Delta C = \frac{\varepsilon \varepsilon_0 A}{\Delta h}. \tag{5.2}$$

When the plate moves horizontally with a distance Δx,

$$\Delta C = \frac{\varepsilon \varepsilon_0 L_x \Delta x}{h}, \tag{5.3}$$

where L_x is the plate length. For example, for air between two plates with a length of 100 μm and a gap distance of 1 μm, a 10 μm lateral displacement would create 8.85×10^{-15} C capacitance change. To increase such a capacitance change value for a better sensor sensitivity, a comb-drive sensor design can be implemented [197], where N fingers could move laterally with respect to each other with a distance Δx, and could increase the capacitance change by $2N$.

Issues of capacitive sensors include their sensitivity to environmental electromagnetic coupling (electrical shielding could reduce it) and orientational alignment sensitivity to the parallelism of the plates. Their resolution could be very high. The measurement range is typically limited by the motion range of the mobile plate. They are highly linear and have a high bandwidth to detect low- and high-frequency parameters. They are compatible to vacuum. As an example capacitive microsensor, a comb-drive microsensor was used to measure the tens of μN aerodynamic flight forces of a fruit fly [198] with a sensitivity of 1.35 mV/μN, linearity of <4%, and bandwidth of 7.8 kHz. The most common use of comb-drive capacitive sensors is in commercial microaccelerometers, where a proof mass attached to the comb-drive sensor is vibrated using capacitive actuation and the changes of the oscillations due to inertial forces are detected by the sensor. Such an accelerometer can have a sensitivity of 38 mV/G, a range of ±50 G, and a noise of 1 mG/$\sqrt{Hz}$.

5.2.2 Piezoresistive sensing

Piezoresistivity is the change in a materials resistance resulting from a change in stress in the material. The word piezo is derived from the Greek word *piezein*, which means to press tight or squeeze. Metals, semiconductors, and many other materials exhibit a piezoresistive effect. It is typically quantified by the gauge factor G, which is the change in resistance per given strain per resistance:

$$G = \frac{\Delta R}{R\varepsilon_s}, \tag{5.4}$$

where R is the resistance of the material and ε_s is the applied strain. The resistance change ΔR can happen due to geometric or conductivity change in the material. In a rectangular cross-section metal element with the length, width, and thickness dimensions L, w, and t, respectively, on a microrobot, the material's resistance is $R = \rho L/A$, where the cross-sectional area $A = wt$. The change in the resistance dR/R can be computed as

$$\frac{dR}{R} = \frac{dL}{L} + \frac{d\rho}{\rho} - \frac{dA}{A} \tag{5.5}$$

$$= (1 + 2\nu)\varepsilon_s + \frac{d\rho}{\rho}, \tag{5.6}$$

where $\varepsilon_s = dL/L$ and $dA/A = -2\nu\varepsilon_s$. Thus,

$$G = 1 + 2\nu + \frac{d\rho}{\rho\varepsilon_s}. \tag{5.7}$$

For metal elements, $d\rho/(\rho\varepsilon_s)$ term is negligible, and the geometric change is the dominant mechanism to change the element's resistance, which results in $G \approx 1 + 2\nu$. In metals, the highest G value is 1.8 (i.e., $\Delta R/R$=1-2%) assuming the highest ν is 0.4 for a metal and the highest $\varepsilon_s = \Delta L/L$ is 1%, close to the metal's yield strength.

For semiconductor piezoresistive elements such as Boron or Phosphor ion-doped silicon and polysilicon, the conductance change-based piezoresistive effect is dominant so that $G \approx \pi_L E$, where $\pi_L = d\rho/(\rho\varepsilon_s E)$ is the piezoresistive coefficient and E is the material Young's modulus. π_L depends on the crystallographic orientation because silicon and polysilicon are anisotropic crystalline materials.

P-type and n-type ion-doped silicon piezoresistive elements can have $\pi_{L<111>} = 93.5\times10^{-11}$ Pa^{-1} and $\pi_{L<100>} = -102.2\times10^{-11}$ Pa^{-1}, respectively. For p-type $Si_{<110>}$, the resultant $G = 133$ is possible, which is two orders of magnitude larger than metal sensor elements. Doped silicon piezoresistive elements can be easily integrated to a microrobot body during a surface or bulk micromachining process by doping the selected element areas with Boron or Phosphor ions at a rate of 10^{15} atoms/cm^3 using a mask. Thus, a large number of, highly integrated piezoresistive microsensors can be fabricated in a wafer-level process. Such sensor elements require conductive (e.g., aluminum) electrical lines patterned around them with a proper electrical isolation layer (e.g., Si_3N_4) to prevent any current leakage, especially for a microrobot operating in ionic liquids.

Polysilicon piezoresistive elements can have G values -40 and 20 for p-doped and n-doped cases, respectively. They have lower G values than doped silicon elements while they have less current leakage issues and can have larger voltages for sensing. Such elements can be micropatterned using photolithography on microrobot surfaces after depositing a polysilicon layer at 560°C using a low-pressure chemical vapor deposition (LPCVD) process and annealing at 1,000-1,100°C. These elements also need to be doped with ions at a higher rate of 10^{19} atoms/cm^3 to increase G.

A Wheatstone bridge and a signal amplifier are used to convert the resistance change in the piezoresistive elements into an electrical signal (voltage difference). Such simple electronics is possible to integrate to the sensor element and microrobot using CMOS microfabrication processes. Such on-chip electronics can reduce the electrical noise significantly. For example, the contractions of a rat heart cell were measured in response to different chemical stimuli using an on-chip CMOS electronics integrated piezoresistive sensor on a silicon microgripper, where such a sensor had 10:1 SNR [199].

The major issues of piezoresistive sensors are electrical noise and thermal drift. Due to possible changes in the environment temperature, resistance of piezoresistive sensors could also change by time. The electrical noise limits the resolution of such sensors, and their noise could be minimized by proper electrical shielding or on-chip CMOS electronics. However, thermal drift is a more difficult problem. As a possible solution to minimize thermal drift, a dummy sensor can be fabricated where a differential signal reading between the real and the dummy sensor could remove the temperature change effects. Moreover, if the application allows, the sensor region could be immersed to a liquid or covered with a thermal insulation layer to provide a constant temperature for the sensor element.

Piezoresistive sensors could be used to design a variety of compact single or an array of deflection, strain, force, pressure, flow, accelerometer, gyroscope, biochemical, etc. sensors [200–202] for a given microrobot application. For example, a silicon microcantilever with a piezoresistive sensing element at the base could detect any molecular binding event on its surface if its surface is coated with the proper specific chemical coating [203]. Such detection is possible because the binding event induces stress on the microcantilever, generating strain and stress at its base. Such biochemical sensors could be compact and highly integrated.

As another important advantage of piezoresistive sensors, they can be fabricated as sensor arrays integrated to a microrobot where their electrical signal detection can be easily decoupled and integrated using CMOS electronics. Here, such a decoupled sensor array could enable detection of flows with a high spatial resolution or simultaneous detection of different biochemical agents or molecules.

5.2.3 Optical sensing

Because optical sensing requires an aligned and focused light source, such as a lase diode or a vertical cavity surface emitting laser (VCSEL), a photodetector, and processing electronics, it is challenging to design and fabricate an on-board optical sensor with all components at the sub-millimeter scale for a microrobot. Therefore, all current optical microsensors use remote lighting and photodetection to sense at the small scale remotely. In this case, the light source needs to be able to penetrate the microrobot workspace and needs to be focused on the microrobot all the time, which could be limiting factors for a given application. For example, the human body is not transparent to visible and UV light; it is semi-transparent (until around 10-20-mm-deep regions under the skin) to near infrared (NIR) light and transparent to x-ray. In contrast, for microfluidic biological applications, the robot workspace could be transparent to visible and UV light, which enables many remote optical sensing techniques for the robot and its environment. For example, a nickel-based magnetic swimming microrobot could be used to sense environmental oxygen in water by coating it with an iridium phosphorescent complex in a polystyrene matrix [204], which phosphoresces in the presence of oxygen when excited by a blue light-emitting diode. After excitation, the luminescent lifetime changes as a function of the oxygen concentration around the microrobot. Through optical detection, a measurement is made; as a result, this sensing mechanism requires line-of-sight, unobscured monitoring of the microrobot in its environment, which restricts the types of measurement tasks and environments in which the microrobot may perform.

5.2.4 Magnetoelastic remote sensing

Magnetoelastic materials can be employed as sensors on microrobots to measure properties of an environment [175, 205]. The sensing element itself is a free-hanging beam constructed out of a magnetostrictive material, which vibrates in the longitudinal direction after activation. When excited by a remote electromagnetic, acoustic, or optical signal, the magnetoelastic sensor will respond with a secondary decaying electromagnetic signal that can be measured using an external sensing system (see Figure 5.2 for an electromagnetic signal-based sensing scheme). This signal will change as the mass loading or elastic properties of the magnetoelastic sensor are varied. By binding probe molecules onto the sensor (in a similar fashion to cantilever beams), this sensor can be used to detect temperature [206], humidity, flow velocity, liquid viscosity and density [207], pH [176], magnetic field strength, and chemical or gas concentrations [175]. This approach is particularly advantageous because the sensor can be read using electromagnetic signals; as a result, the sensor does not need to be visible to be read and can operate in closed or opaque environments. Moreover, such a remote microsensor could be easily integrated into mobile microrobots with no need of on-board measurement electronics, data transfer, and power.

The magnetoelastic sensor operates on the principle of magnetostrictively induced vibrations. For a thin ribbon of material length l, the first resonant frequency f_0 for longitudinal vibrations can be approximated by [175]

$$f_0 = \frac{1}{2l}\sqrt{\frac{E_m}{\rho_m(1-\nu_m^2)}}, \tag{5.8}$$

where E_m is the Young's modulus of the magnetoelastic material, ρ_m is its density, and ν_m is its Poisson's ratio.

Many of the magnetoelastic sensors used in the literature require coating the magnetoelastic material with an active chemical layer, which changes the material's effective Young's modulus (E_{eff}) and density (ρ_{eff}):

$$E_{eff} = \alpha_m E_m + \alpha_c E_c, \tag{5.9}$$

$$\rho_{eff} = \alpha_m \rho_m + \alpha_c \rho_c, \tag{5.10}$$

where α_m and α_c are the fractional thicknesses of the magnetoelastic material and the coating, respectively.

Once f_0 of a magnetoelastic sensor is determined, changes in its natural frequency can arise due to changes in the sensor, such as added mass due to target

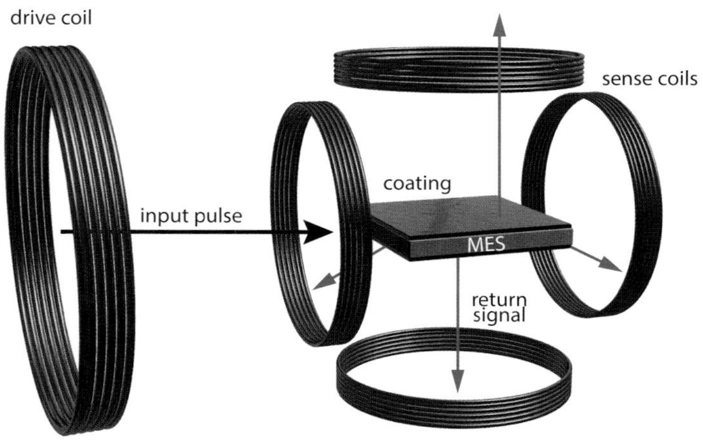

Figure 5.2

Schematic of a possible sensing scheme used for detecting changes in natural frequency for a magnetoelastic sensor (MES). An electromagnetic pulse can be sent using the drive coil, causing the magnetoelastic sensing element to oscillate. This oscillation would create a decaying electromagnetic signal, which could be picked up by the sensing coils. The electromagnetic response could be analyzed to determine the amplitude and resonant frequency of response.

molecules adhering to the device. For mass increases Δm that are much smaller than the sensor mass m (i.e., $\Delta m/m \ll 1$), and which have negligible effect on the sensor's stiffness, the change in resonant frequency, Δf, is:

$$\Delta f = -f_0 \frac{\Delta m}{2m}. \tag{5.11}$$

When operating within a liquid, the liquid's density and viscosity act to dampen vibrations, decreasing the amplitude of the response, and also adding an effective mass to the sensor, thereby decreasing its resonant frequency. The high viscosity limit for determining the theoretical response of a magnetoelastic sensor in liquid is described in [207]. The figure of merit for determining which case is more appropriate is the penetration depth of the waves produced by the sensor oscillations, δ, compared with the characteristic length scale of the volume of fluid, $2h$. Explicitly:

$$\delta \approx \sqrt{\frac{\eta}{\pi \rho_l f_0}} \ll 2h \quad \text{for low viscosity,} \tag{5.12}$$

$$\delta \approx \sqrt{\frac{\eta}{\pi \rho_l f_0}} \gg 2h \quad \text{for high viscosity,} \tag{5.13}$$

where η is the fluid's dynamic viscosity and ρ_l is the fluid density.

For low-viscosity fluids, the change in resonant frequency is:

$$\Delta f = -\frac{\sqrt{\pi f_0}}{2\pi \rho_m d}\sqrt{\eta \rho_l}, \tag{5.14}$$

where d is the thickness of the sensor. For high-viscosity fluids, the change of resonant frequency is:

$$\Delta f = -\frac{1}{3} f_0 \frac{\rho_l h}{\rho_m d}. \tag{5.15}$$

With magnetostrictive materials, the effective modulus of elasticity is a function of the material's magnetic anisotropy. Changes in applied magnetic field (H) change the material's magnetization (M) and its permeability (χ), which both in turn affect the relationship between internal stress and strain. This results in a field-dependent effective Young's modulus, and hence a field-dependent resonant frequency, $f(H)$:

$$f(H) = f_0\left(1 + \frac{9E_m \lambda_s^2 M^2}{M_s^2}\chi\right), \tag{5.16}$$

where λ_s is the saturation magnetostriction coefficient of the material and M_s is the saturation magnetization of the material. The dependence of M on H is a function of both the sensor geometry and material, whereas χ, λ_s, and M_s are all functions of temperature in addition to intrinsic material properties. These dependencies can be advantages, in that it is possible to cancel out temperature effects by choosing an appropriate magnetic bias fields, thereby allowing the measurement of some other parameter in an environment where temperature is difficult or impossible to control. Alternatively, an appropriate bias field can be used to change the resonant frequency, if desired, or enhance the sensitivity to temperature, if that is the parameter being measured.

6 On-Board Actuation Methods for Microrobots

Microrobots can be actuated using on-board microactuators, self-propelled using physical or chemical interactions with their operation medium or biological cells attached to them, or remotely actuated. In this and the next two chapters, each of these actuation methods is explained in detail. This chapter covers the commonly used on-board actuation methods, such as piezoelectric, shape memory material, electroactive polymer, and MEMS-based electrostatic, capacitive, and thermal actuators, which could be scaled down to tens or hundreds of micron scale and integrated into mobile microrobots. However, for such actuators, we need an on-board electrical energy source (such energy could be also remotely transferred), driving electronics, processor, and controller, which make them more challenging to be used for mobile microrobots due to current technological miniaturization limits on such on-board components. Therefore, many example cases of these actuation techniques will typically be given for mobile millirobots.

6.1 Piezoelectric Actuation

Piezoelectric actuation is one of the most promising on-board milli/microscale actuation methods due to its high force output, high power density, high bandwidth (up to MHz), small form factor (can be integrated into robotic structures as thin/thick films), and low power consumption (tens of μW or mW). However, it provides small strain (less than 0.1% typically and up to 1% for single crystal piezos only); has some non-linearities, such as hysteresis, creep, softening, and drift; requires hundreds of volts with voltage amplifiers; may break, age, or depolarize; and is not biocompatible if not coated properly.

The piezolectric effect was discovered in 1880 by Jacques and Pierre Curie. The word *piezo* means *press* in Greek; thus, *piezoelectric* means mechanical pressure inducing electric polarization on a piezo material (pressure-sensing mode). Piezo materials can also generate mechanical stress/strain by an electric field input (actuation mode). Therefore, they can be used as both actuators and sensors at the same time.

There are a wide range of piezo actuator materials, such as polar non-ferroelectric single crystals (ZnO, AlN, tourmaline, Rochelle salt, quartz) with domain walls but with no grain boundaries, ferroelectric polycrystals ($BaTiO_3$, $Pb(Zr\text{-}Ti)O_3$, $LiNbO_3$, $LiTaO_3$, etc.) with both grain boundaries and domain walls, ferroelectric single crystals (PZN-PT, PMN-PT, etc.), some natural organic substances (rubber, wool, hair, silk, etc.), and specific thermoplastic fluoropolymers (polyvinylidene fluoride (PVDF)). The most typical commercial piezo materials are PZT (i.e., $Pb(Zr_x\text{-}Ti_{1-x})O_3$, lead zirconate titanate, $0 \leq x \leq 1$) ferroelectric polycrystals, which could have soft (e.g., PZT-5H) and hard (e.g., PZT-5A) types, optimized for more strain less force and less strain but more force, respectively.

Polycrystalline ferroelectric materials, such as PZT have a perovskite crystal structure with a general formula of $A^{2+}B^{4+}O_3^{2-}$. Such crystals have a tetragonal symmetry and inherently built-in electric dipoles at temperatures less than the Curie temperature (T_c). At temperatures higher than T_c, such crystals have cubic symmetry and no electric dipoles (paraelectric mode). At $T < T_c$, if the piezo material is not poled, its electric dipoles are randomly oriented. In this form, the piezo material can not be used as an actuator because the mechanical stress due to a constant electric field direction would cancel each strain from the random dipoles, inducing negligible total strain. Therefore, the piezo material must be poled before using it as an actuator, which is achieved by applying a high electric field at a temperature slightly smaller than T_c. During such a poling process, electric dipoles all align in the electric field direction, and when the field is turned off, they can still preserve their aligned dipole directions.

When a voltage V is applied to the piezo material, it has a mechanical motion (contraction/expansion) δ, which can be approximated by a linear relation of

$$\delta = d_{33} V, \tag{6.1}$$

where d_{33} is a piezoelectric constant. However, this relation is not accurate because of the many non-linear effects such as:

- Hysteresis (a retarded reorientation of dipole domains) occurs during actuation, which induces non-linearity and energy losses in the piezo material.
- Creep (shift in strain after motion) can occur after changing the driving field and then holding it constant, where more and more dipoles can orient themselves in the applied electric field due to mutual influence.
- At high electric fields, the piezo material can change its stiffness (soften).

In addition to possible non-linearities, if no external field is applied for a long time, then some domains can lose their polarity; repoling could be needed. Micro-cracking due to high mechanical and electrical loading and driving in the bipolar mode could reduce the lifetime (durability) of piezo actuators. When such high loading is minimized and they are driven in the unipolar mode,

Table 6.1
Physical properties of a hard PZT, optimized for higher force and lower strain applications, ceramic actuator material (PZT-5A) ($\varepsilon_0 = 8.854 \times 10^{-12}$ F/m)

Parameter	Value
Curie temperature, T_c (°C)	220
Density, ρ (kg/m^3)	7,900
Frequency constant, N (Hz.m)	2,000
Compressive strength (MPa)	600
Tensile strength (MPa)	50
Relative dielectric constant, $\varepsilon_r/\varepsilon_0$	1,700
d_{31} charge constant (C/N)	-171×10^{-12}
d_{33} charge constant (C/N)	374×10^{-12}
Coupling factor, k_{31}	0.62
Poisson ratio, ν	0.3
Young's modulus, E (GPa)	70

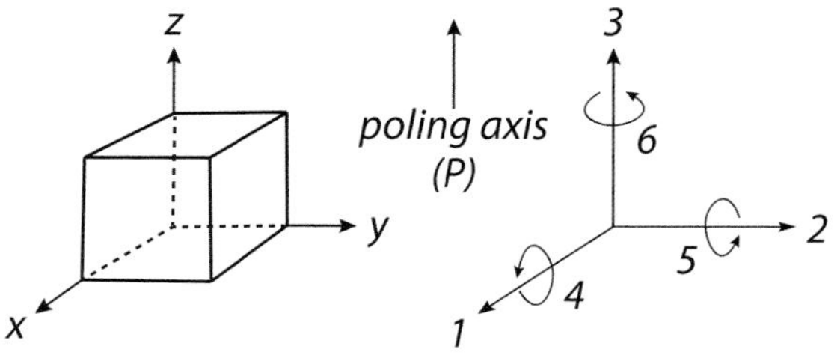

Figure 6.1
Basic directions in a 3D anisotropic polycrystalline piezo ceramics.

they could have 10^9 cycles of lifetime, which is sufficient for a large number of robotic applications. Piezo materials have high compressive strength; therefore, they can sustain compressive loads up to hundreds of MPa. However, their tensile and shear strengths are low, which make them fragile/brittle against tensile and shear loads. To minimize such brittle behavior, pre-stressing (preloading the piezo material externally or internally by an external force, an internal thermal mismatch, or higher electric fields) can be utilized.

For example commercial hard piezo material properties are listed in Table 6.1. Piezoelectric charge constant d_{ij} is the electric polarization generated due to the mechanical stress, where i denotes the direction of applied electric field and j denotes the direction of induced strain. Thus, d_{31} is the constant for the

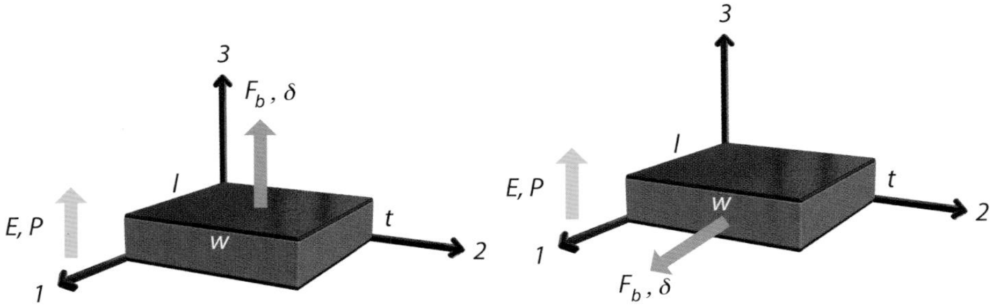

Figure 6.2
(a) Axial (left image) and (b) transversal (right image) type piezo actuation.

strain in x-direction induced by the electric field in z-direction. Such index numbering is related to the 3D anisotropic polycrystalline directions of the piezo ceramic material, as shown in Figure 6.1.

Piezo actuators can be designed in many different ways, such as axial (d_{33} mode), transversal (d_{31} mode), composite (multi-layer/stack), and flexural, where each design type gives a different force (blocking force, F_b) and displacement (δ) output in a specific direction.

In the axial type (longitudinal effect),

$$\delta = d_{33} V \tag{6.2}$$

$$F_b = \frac{lw}{s_{33}^E t} d_{33} V \tag{6.3}$$

$$K_m = \frac{F_b}{\delta} = \frac{lw}{s_{33}^E t} \tag{6.4}$$

$$f_r = \frac{N_3^D}{t}, \tag{6.5}$$

$$\tag{6.6}$$

where Figure 6.2 shows l, w, and t dimensions of a rectangular piezo material polarized (P) in z-direction. Electric field E_3 is applied in z-direction, which induces F_b and δ in z-direction. Here, K_m is the actuator mechanical stiffness, f_r is its resonant frequency (first mode) when the base of the actuator is not fixed (when the base is fixed, the denominator becomes $2t$), N_3^D is the frequency constant, and $s_{33}^E = 1/E$ is the compliance, where E is the anisotropic ceramic Young's modulus in z-direction.

In the transversal type (longitudinal effect), electric field E is applied in z-direction, which induces F_b and δ in x-direction. Here,

$$\delta = -\frac{l}{t} d_{31} V \tag{6.7}$$

$$F_b = -\frac{w}{s_{11}^E} d_{31} V \tag{6.8}$$

$$K_m = \frac{hw}{s_{11}^E l} \tag{6.9}$$

$$f_r = \frac{N_1^E}{l}, \tag{6.10}$$

where the negative sign is due to the reduction in length. This is the most typical piezo actuation type for a stack piezo.

High voltages are needed for high stroke for the axial and transversal piezo actuators. To lower the applied voltage, n number of thin piezo layers can be stacked on top of each other to create a composite stack design. Thus, using the same applied voltage V, you can get n times more displacement. Such composite stacks in the form of a piezo tube or other designs with 3D (xyz) translational motion capability are frequently used in high-resolution atomic scale nanopositioning systems, with few or tens of μm motion range for scanning force microscopy.

Although above three piezo actuation methods provide high output forces, they have limited range of motion, which limits their use for robotic applications. However, flexural-type piezos can increase range of motion significantly by reducing the stiffness of the actuator. Typical flexural-type piezos can be a unimorph (one active piezo layer) or bimorph (two active piezo layers) type, where the piezo layer is bonded to a passive elastic layer.

6.1.1 Unimorph piezo actuators

A standard rectangular cross-section unimorph actuator under activation is illustrated in Figure 6.3. The actuator consists of a single piezoelectric layer bonded to an elastic layer. Steel or titanium is usually chosen for the elastic layer. When a voltage is applied across the thickness of the piezoelectric layer, longitudinal and transverse strains develop. The elastic layer opposes the transverse strain, which leads to a bending deformation.

For a one end fixed and one end freely deflecting unimorph actuator, as in Figure 6.3, the tip displacement δ, F_b, f_r, K_m, and mechanical quality factor Q

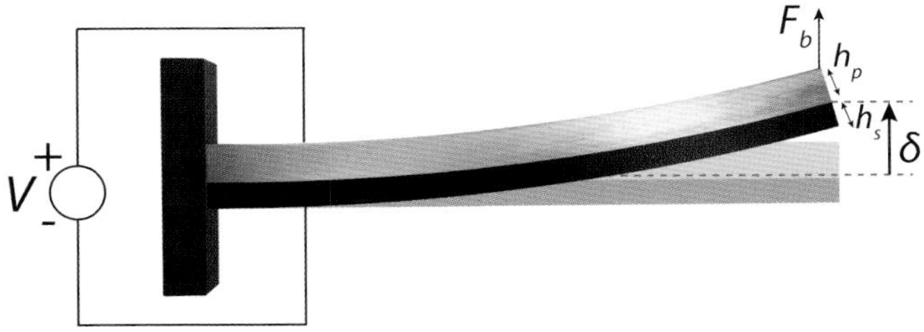

Figure 6.3
Basic cantilevered rectangular-shaped piezoelectric unimorph actuator design for small-scale actuation.

Table 6.2
PZT-5H, PZN-PT, and steel layer properties

	PZT-5H	PZN-PT	Steel
E (GPa)	61	15	193
ρ (kg/m^3)	7,500	8,000	7,872
d_{31} (C/N)	320×10^{-12}	950×10^{-12}	N/A
k_{31}	0.44	0.5	N/A
K_3	3,800	5,000	N/A
E_3 (V/m)	1.5×10^6	10×10^6	N/A

can be written as [208]:

$$\delta = \frac{3l^2}{h_p^2}\frac{AB(B+1)}{D}d_{31}V \tag{6.11}$$

$$F_b = \frac{3wh_p}{4s_pl}\frac{AB(B+1)}{AB+1}d_{31}V \tag{6.12}$$

$$f_r = \frac{\lambda_i^2 h_p}{4\pi l^2}\sqrt{\frac{E_p}{3\rho_p}\frac{D}{(BC+1)(AB+1)}} \tag{6.13}$$

$$Q = \frac{f_r}{f_{r1}-f_{r2}} \tag{6.14}$$

$$K_m = \frac{F_b}{\delta} = \frac{wh_p^3}{4s_pl^3}\frac{D}{AB+1}, \tag{6.15}$$

where $A = s_p/s_s = E_s/E_p$, $B = h_s/h_p$, $C = \rho_s/\rho_p$, $D = A^2B^4 + 2A(2B + 3B^2 + 2B^3) + 1$. Here, f_{r1} and f_{r2} are the frequencies, where the deflection magnitude drops to 0.707 of its resonance peak value. $s_p = 1/E_p$ and $s_s = 1/E_s$

are the elastic compliances, h_p and h_s are the thicknesses, E_p and E_s are the Young's moduli, and ρ_p and ρ_s are the densities of the piezoelectric and steel layers, respectively, and λ_i is the eigenvalue, where i denotes the resonance mode, i.e., first mode $\lambda_1 = 1.875$, and the second mode $\lambda_2 = 4.694$. For the PZT-5H, PZN-PT, and steel layers, Young's modulus, density, d_{31}, coupling factor k_{31}, relative dielectric constant $K_3 = \varepsilon/\varepsilon_0$, and maximum electric field E_3 values are given in Table 6.2.

At resonance, the actuator would oscillate with an amplitude of δ_r assuming a linear behavior, where Q can also be defined as

$$Q = \frac{m^* \omega_0}{b_a} = \frac{\delta_r}{\delta}, \tag{6.16}$$

assuming a second-order linear model, where m^* is the effective mass of the unimorph bending actuator ($m^* \approx 0.24 m_a$ for a one end fixed cantilever beam, where m_a is the unimorph mass), $\omega_0 = 2\pi f_r$, and b_a is the actuator damping.

Fixing the PZT-5H piezo layer dimensions and applied voltage (h_p = 127 μm, l = 16 mm, w = 3 mm, and V = 150 V), the elastic layer's thickness h_s can be tuned to optimize the displacement or mechanical energy output of the unimorph actuator. As can be seen in Figure 6.4, if δ deflection is to be maximized, then h_s should be around 20 μm. However, typical robotic applications require maximizing the mechanical energy output, which can be achieved by choosing h_s as 56 μm in this case.

6.1.2 Case study: Flapping wings-based small-scale flying robot actuation

As an example case study, this section covers the design of unimorph piezo actuators for a microaerial flapping wing mechanism to be able to lift off. PZT-5H and PZN-PT are investigated as piezoelectric layers in the unimorph actuators. Design issues for microaerial flapping actuators are discussed, and the desired parameters of the unimorph actuators are determined.

Flapping mechanisms require actuators with a large periodic stroke (rotational) motion (30-150°) at a high speed (10-100s of Hz) with large output forces for overcoming the aerodynamic damping. Moreover, light weight (10s of mg), high efficiency, long life time, and compact size are important issues. Piezoelectric actuators with proper design almost satisfy all of these requirements. Flexural bending actuators generate large deflection with low weight. Therefore, bimorphs and unimorphs are more suitable for microaerial flapping wing applications. Because they are easier to fabricate, the unimorph type is selected.

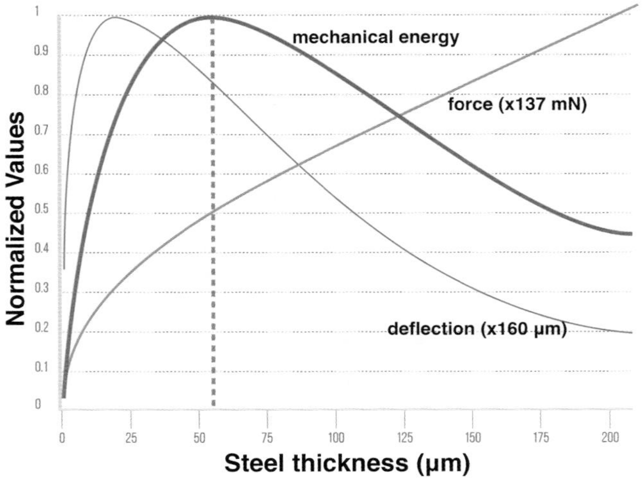

Figure 6.4

Fixing the PZT-5H piezo layer dimensions and applied voltage, the elastic layer's thickness h_s can be tuned to maximize the displacement or mechanical energy output of the unimorph actuator.

For the flapping wing design considerations, the linear motion equations for unimorphs are converted to the rotational motion ones for simplicity. Assuming the actuator tip deflection is small, actuator rotation angle θ, output torque τ_a, and rotational stiffness K_a are given as follows:

$$\theta = \frac{\delta}{l} \tag{6.17}$$

$$\tau_a = F_b l \tag{6.18}$$

$$K_a = K_m l^2 \; . \tag{6.19}$$

Here, the maximum input voltages are $V_{max} = E_3 h_p$, e.g., for h_p = 100 μm, V_{max} = 150 V and V_{max} = 1000 V for PZT-5H and PZN-PT unimorphs, respectively. Moreover, the actuators are driven unipolarly, i.e., $V > 0$, to maximize their life time. Thus, the wing motion is $\phi \in [0 \, , \phi]$ at DC, and $\phi \in [\phi/2 - \phi_r \, , \phi/2 + \phi_r]$ at resonance.

For a flapping mechanism with a wing load on it, design parameters, such as unimorph dimensions, output torque, resonant frequency, required transmission ratio, quality factor, weight, etc. are to be selected for optimal performance. In a possible flapping mechanism, a four-bar transmission mechanism

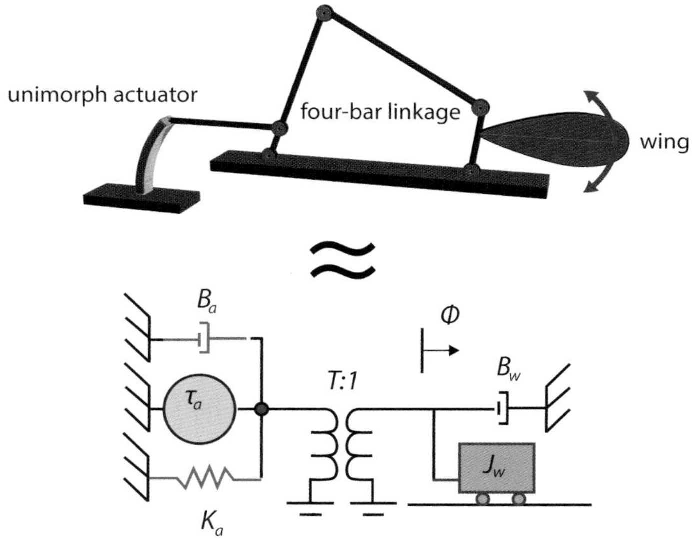

Figure 6.5
Linear dynamic model (bottom image) of a flapping wing design (upper image) with a piezo actuator, lossless transmission, and wing.

can be coupled with the unimorph actuator for stroke amplification [165, 209–211]. The wing induces a load with inertia J_w and aerodynamic damping B_w on the four-bar mechanism with stiffness K_t and stroke amplification (transmission ratio) T. Linear approximate dynamic modeling of such system, as illustrated in Figure 6.5, can be given as

$$J_w\ddot{\phi} + B_w\dot{\phi} + \left(\frac{K_a}{T^2} + K_t\right)\phi = \frac{\tau_a}{T}, \tag{6.20}$$

where ϕ is the flapping stroke angle and K_a is the actuator rotational stiffness. Here, the actuator damping B_a and inertia J_a are assumed to be negligible with respect to the load damping and inertia.

For a given load power requirement, the actuator dimensions are to be chosen for ease of mechanical drive, fabrication, and drive voltage requirements. Considering an insect-inspired flying robot modelled after a blowfly, with mass $m = 0.1$ g, wing beat frequency of $\omega = 2\pi\, 150$ rad/s, and wing stroke amplitude $\phi_r = 70^\circ$ at resonance, the net wing lift force must match the insect weight of 10^{-3} N. Although lift and drag forces are generally proportional to the square of velocity in the quasi-steady state, we choose a linear damper with a force at peak wing velocity equals to the weight of the flying robot as an upper bound. (Note that the linear damper overestimates the damping force for all wing velocities less than the peak velocity). Hence, the wing damping B_w (at the wing hinge) can be estimated from:

$$B_w = \frac{mgl_w}{\omega\phi_r}, \tag{6.21}$$

where m = 0.1 g, g = 9.81 m/s^2, and l_w is the length of the wing center of pressure. For l_w = 10 mm, $B_w = 8.65 \times 10^{-9}$ N.s.m.

Blow flies have a relatively low Q, estimated on the order of 1-3. For the flying robot, we choose the quality factor of the wing and thorax as Q_w = 2.5 because a higher Q_w system requires a lower transmission ratio and less actuator motion at DC. To have a low Q_w, i.e., maneuverable wing, the wing inertia is:

$$J_w = \frac{Q_w B_w}{\omega} = 2.26 \times 10^{11} \text{ kg.m}^2. \tag{6.22}$$

The actuator stiffness, as seen at the wing hinge, must resonate at ω, hence:

$$K_1 = \frac{K_a}{T^2} + K_t = J_w\omega^2 = 2 \times 10^{-5} \text{ N.m}. \tag{6.23}$$

The four-bar transmission converts the small rotation of the actuator θ to the wing rotation ϕ by a transmission ratio T. At DC, the displacement of the wing is

$$\phi = \frac{\tau_a}{TK_1} + K_t = \frac{2\phi_r}{Q_w} = T\theta. \tag{6.24}$$

For a given T and the desired wing flapping amplitude ϕ_r at resonance frequency ω,

$$\tau_a = \frac{2K_1\phi_r T}{Q_w} = \phi K_1 T \tag{6.25}$$

$$\theta = \frac{\phi}{T} = \frac{2\phi_r}{Q_w T}. \tag{6.26}$$

For a given h_s, h_p, and V values, l and w can be computed as

$$l = \frac{h_p^2 D}{3d_{31}AB(B+1)V}\theta \tag{6.27}$$

$$w = \frac{4s_p}{3d_{31}h_p V}\frac{AB+1}{AB(B+1)}\tau_a. \tag{6.28}$$

Table 6.3
Selected flying robot unimorph actuator design parameters for different T and h_s values and unimorph piezo types

Type	T	h_s (μm)	$l \times w$ (mm × mm)	F_b (mN)	δ (μm)	f_r (Hz)	m_a (mg)
PZT-5H	44	76	16 × 2.9	54	354	464	74
PZN-PT	36	76	5 × 1.4	142	135	3,032	12
PZT-5H	39	50	16 × 3.6	49	393	406	78
PZN-PT	28	50	5 × 1.3	109	176	2,548	10

The average power at the wing is also another important parameter for the design, which can be computed from

$$P_w = \frac{\tau_a^2 B_w}{8T^2(B_w + B_a/T^2)^2} = \frac{(mgl_w)^2 B_w}{2(B_w + B_a/T^2)^2}, \tag{6.29}$$

where B_a=$K_a/(Q\omega)$=20 and K_a=$T^2(K_1 - K_t)$.

Furthermore, the mass of the actuator m_a is limited for enabling a total flying robot mass of m = 0.1 g. Therefore, $m_a = (\rho_p h_p + \rho_s h_s)lw$ should also be checked.

Considering the available piezoelectric materials in Table 6.2, V = 150 V and h_p = 127 μm, and V = 250 V and h_p = 136 μm are fixed for the PZT-5H and PZN-PT layers, respectively. Taking $J_w = 2.26\times10^{-11}$ kg.m^2, $B_w = 8.65\times10^{-9}$ N.s.m, $\omega = 2\pi\,150$ rad/s, $\phi_r = 70^\circ$, $Q_w = 2.5$, and $K_t = 5.3\times10^{-6}$ N.m/rad, P_w = 4.7 mW and required l, w, F_b, δ, P_w, f_r, and m_a values are computed for given T and h_s as shown in Table 6.3. From these values, it can be seen that $16\times3\times0.21$ mm^3 size PZT-5H and $5\times1.3\times0.22$ mm^3 size PZN-PT would enable the desired 140° wing flapping at 150 Hz with relatively low masses. To lift off with the PZT-5H unimorph, V should be increased to 250 V, lowering actuator mass to 26 mg per wing.

6.1.3 Bimorph piezo actuators

In the bimorph piezo design, the passive elastic layer in the previous unimorph design is replaced by the same size and material active piezo layer. To be able to create a bending motion, while the upper piezo layer is contracting, the lower one needs to contract and vice versa. Thus, the amount of deflection of the bimorph design can be much larger than the unimorph design with the same

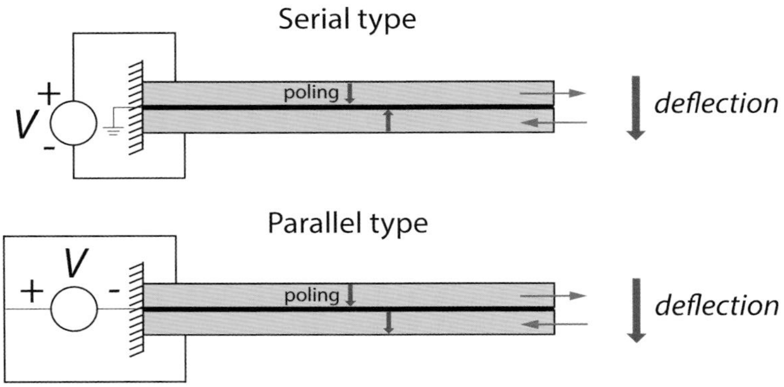

Figure 6.6

Serial and parallel type bimorph piezo actuators, where the two active piezo layers are bonded to each other with the given poling directions (the arrow in each piezo layer) and electrical connections. The upper layer expands and the lower layer contracts for enhanced bending motion.

applied voltage. Electrical connections for each piezo layer could be realized using a serial or parallel type, as illustrated in Figure 6.6. The parallel type has higher sensitivity and less risk of depolarization.

6.1.4 Piezo film actuators

Piezo ceramic or polymer materials, such as ZnO, PZT, PVDF, and AlN, can be deposited, coated, or grown on microrobot and other surfaces as a thin or thick film as the most promising form of piezo actuators to be used for microrobots. When the piezo materials are formed as a film, their material properties can be different from the bulk material. For example, d_{31} value for piezo films is on the order of -(5-100) pC/N and K_3 is 300-1,500, which are lower than the same bulk piezo materials. In addition, piezo films can have many defects and low fabrication yield. Moreover, film bonding to the structure and creating electrical contact to the film electrodes are some practical challenges for each robot material and structure.

Sol-gel deposition is one of the standard, low-cost piezo film fabrication techniques. In this technique, a PZT liquid solution is spin coated on a given substrate, heated at 600°C for 0.1 hour, spin coated, and heated again until reaching the right thickness. As the final step, the film is annealed at 700°C for 1 hour. Electrical poling is needed after the film is formed. Magnetron sputtering for ZnO films, pulsed-laser ablation for PMN-PT films, hydrothermal method for PZT films, and metal-organic chemical vapor deposition for ZnO films are other common piezo film fabrication techniques.

6.1.5 Polymer piezo actuators

Polyvinylidene fluoride (PVDF) piezo films have the advantage of being flexible, not brittle, lightweight (low density), and higher strain due to lower stiffness. Typical values of a PVDF piezo material are E=2 GPa, ρ = 1,780 kg/m^3, K_3 = 1200, and d_{31} = -20 pC/N. As drawbacks, its d_{31} value is almost a magnitude order lower than the piezo ceramic materials, which reduces their force and strain output significantly, and it has viscoelastic dynamics and high creep behavior compared with piezo ceramics. Polymer piezos are mainly used in active suspension, active noise control, and acoustic actuator applications, and they could be used as a microrobot actuator if coated as a thin film onto the robot structures.

6.1.6 Piezo fiber composite actuators

A flat monolayer of aligned single- or poly-crystal piezoelectric fibers with 1-10 mm lengths and 5-250 µm diameters have the advantage of being flexible and creating aligned piezo fiber actuators with directional stiffness properties. In some robotic applications, the actuator is desired to have an anisotropic compliance (e.g., low stiffness in specific translational or rotational degrees of freedom and high stiffness in others) for optimal performance and lifetime, such as in the biological actuators and structures. In such cases, piezo fibers are desirable to create such compliance anisotropy.

6.1.7 Impact drive mechanism using piezo actuators

Due to high force output and speed of piezo ceramic actuators, novel actuation mechanisms, such as impact drive mechanism can be used to move a robot or device on surfaces using friction at the small scale. Attaching one end of a piezo ceramics to a main body and the other end to a load with an inertial weight, slow contraction (inertial force $<$ static friction), sudden stop, and fast extension (inertial force $>$ friction) and of the piezo element can induce inertial force on the main body to propel it on the surface with a given speed. Konica used such an actuation mechanism to correct shaking of cameras. Such a simple mechanism can create high-resolution (nanometer scale) positioning systems with long range and high speed. No energy is required for keeping the body's constant position. If the body has enough high friction with its substrate, then it could even climb a wall.

A similar principle is a stick-slip actuation mechanism, where an inertial mass can be guided by deformable piezo legs. If a triangular waveform is applied to bend the piezo legs asymmetrically (fast in one direction and slow in the other direction), then a net inertial force can be induced at the leg tip to move in a direction. In the slow leg deflection case the leg tip sticks to the surface, whereas in the fast deflection case it slips from the surface. Such a stick-slip motion can be coarse, long range, and high speed. For short-range fine positioning, piezo legs can be deformed with a symmetrical wave form, such as a sine wave, which would not induce any net inertial force. Such a mechanism can be used to create miniature robots that could crawl on a surface with high precision, high speed, and long range.

6.1.8 Ultrasonic piezo motors

If a piezo ceramic is formed into an annular disk, then a traveling wave at ultrasonic frequencies (more than 40 KHz) can be induced on its surface by patterning its top electrode in specific short segments and exciting each segment at resonance with a specific waveform. Such a traveling wave-generating disk can be called a stator. Points on the surface of the stator move in retrograde elliptical motions. If the stator is coupled to an annular rotor disk using surface friction and vertical loading, then the stator can rotate the rotor with almost no noise. Such motors can be scaled down to a few millimeter scale and are promising for small-scale robots. Their quiet, fast, and high output and stall torque operation is unique compared with existing tiny electrical motors. They have been used in small-scale flying miniature robots with rotary wings, crawling robots [212], and medical robots [213].

Ultrasonic frequency piezo ceramic vibrations can induce translational linear motion by rotating a treaded screw, as in the case of commercial Squiggle motors. Such piezo motors have few nanometer resolution, are fast (linear speed of 1 μm/s - 10 mm/s), have high force output (up to 5 N), and are tiny (as small as $1.5 \times 1.5 \times 6$ mm^3), quiet, and smooth.

6.1.9 Piezoelectric materials as sensors

Piezo materials can also be used as sensors by measuring their electrical polarization due to exerted mechanical stress on them. Piezo materials have high capacitance due to their high dielectric constant (K_3), which makes quasi-static (<1 Hz) mechanical deflection/stress detection not possible due to charge leakage issues. However, cyclic stresses/bending oscillations more than 1 Hz can be easily detected with them. Such a sensing method is possible in two ways. First, the same piezo material on a given robotic mechanism can be used as both actuator and sensor. In this case, the electrical circuitry needs to be designed to both apply voltage and detect accumulated charge between the two electrodes of the piezo material, and any coupling between actuation and sensing modes needs to be avoided. Next, separate piezo segments or actuators can be used as actuators and sensors, where it is much easier to decouple actuation and sensing signals.

6.2 Shape Memory Materials-Based Actuation

Shape memory alloy (SMA) actuators are commonly used in small-scale robotics due to their compact size, easy operation, high power density, mechanical robustness, and light weight. They contract their length by applied heat and can provide high output forces and power densities like piezos and high strains up to 3% to 4%, which is much higher than standard piezos. They are biocompatible and can be compact as milli/microscale wires, springs, sheets, or films. However, they are slow, not power efficient (around 2% or so due to dissipated heat to the environment), not so durable (around 10^{4-5} cycles), and non-linear due to hysteresis.

Shape memory effect is the ability of certain materials to annihilate a deformation and to recover a predefined or imprinted shape, which is based on a solid-solid phase transition of an SMA within a specific temperature interval. SMAs have two distinct solid phases. In the austenitic state (heated above its austenitic temperature), they have a symmetric β-phase crystal structure with high elastic modulus. In the martensite state (cooled down lower than its martensite temperature), they are twinned and have low elastic modulus. In this state, the material can be de-twinned by external mechanical stress/deformation. If the material is heated more than its austenitic temperature after the mechanical deformation, then it would remember and go back to its austenitic state and recovers its original shape. The first martensitic transformation was observed in 1951 for Au-Cd alloys.

An SMA can be used as an actuator in two ways. First, in the *one-way* memory effect, shape recovery occurs only when the SMA element is heated. In martensitic (cold) state, SMA element is de-twinned and plastically deformed by stretching. Upon heating above its austenitic temperature, it returns back to the non-stretched form. Such an effect can be used to create SMA fasteners, clamps, and coupling sleeves. For example, in two separated bone regions after a fracture, an SMA clamp can be inserted into each bone region in the cold state, and then heated up to contract and clamp two bone regions to each other. SMAs can have up to 8% strain when one-way memory effect is used. Second, in the *two-way* memory effect, shape change occurs upon both heating and cooling repeatedly. SMA is trained to remember both high- and low-temperature shapes. Heat treatment and mechanical training methods are used to create the two-way shape memory effect on an SMA. The heated shape is imprinted by annealing the SMA element at 500 °C or so for a specific duration and cycle time (e.g., 20-100 cycles). In this case, SMAs can have up to 3% to 4% strain.

Two-way SMA shape memory effect is utilized for robotic applications. SMA element is heated above its austenitic temperature to contract and recovers back its original length and shape after cooling down. For thermal activation of SMAs, electrical or photothermal heating can be used. In the former, the Joule heating is used by passing electrical current through the SMA element. This is the most compact and efficient heating method. In the latter, a focused laser beam can be used to heat SMA locally and remotely for untethered microrobotic applications with less efficiency. Moreover, overall surrounding liquid or air medium of the SMA element can be heated, which is typically a bulky and inefficient heating method for mobile robotic applications.

Typical SMA actuator materials are NiTi (nitinol), CuZnAl, and FeNiCoTi. Transformation temperatures of these materials should be high enough for unwanted activation by warm ambient air. Their heating is quick by applying current pulses. However, cooling is slow, influenced by the air or water medium (much faster in water) and size scale of the SMA element (smaller size, i.e., higher surface-to-volume ratio, for fast cooling). Thus, SMA films that can be used in microrobotic applications would have a fast response time (bandwidth) due to their small size scale. Moreover, the same SMA element can be used as a temperature sensor (medium heating can trigger the SMA element contraction) or a strain sensor (measured electrical resistance of the SMA element is a function of its length change).

SMAs can be directly powered by a battery because they can operate with low voltages. However, they need high electrical currents (0.01-1 A), which may require H-bridge current amplifiers, and consume a lot of electrical power in a short time due to their low energy efficiency. They have been used in medical robotics (e.g., flexible active catheters), space robotics, robotic clutches, insect-inspired robots, etc.

Shape memory polymers are also becoming an interesting actuation material candidate due to their large strains and flexibility [214]. They can also be trained to recover specific shapes when they are heated, but current shape memory polymer actuators are not repeatable enough yet (only a few cycles are possible). When their repeatability is improved, they would be able to be used in microrobotic applications.

6.3 Polymer Actuators

A wide range of polymer materials can be used for small-scale actuation. We can group them into three main categories: magnetically and thermally actuated and electroactive polymer actuators. First, magnetic micro/nanoparticles can be embedded inside or magnetic thin films can be coated on various polymer materials to actuate them with remote magnetic fields or gradients. For example, as shown in Figure 4.5 [25], soft elastomeric polymers can be embedded with hard NdFeBr magnetic microparticles, which could actuate an undulating milli/microswimmer. Such a soft magnetic polymer actuation concept has a significant potential for complex and dynamic actuation of mobile microrobots remotely. Next, shape memory polymers and other temperature-sensitive polymers can be heated locally using photothermal or electrical Joule heating methods to create actuation for microrobots, such as in Section 6.2. Finally, electrically active polymer (EAP) actuators, which can contract or expand in response to electric energy, are the most common polymer actuators for small-scale robotics.

EAPs have two main subgroups: ionic and electronic EAPs. Ionic EAPs, such as conductive polymer, ionic polymer-metal composite, and ionic gel actuators, involve mobility or diffusion of ions (mass transport), are typically wet, require low driving voltage (1-10 V), and yield large displacements and low forces. Electronic EAPs, such as dielectric elastomer, electrostrictive polymer, electro-viscoelastic elastomer, ferroelectric polymer, and liquid crystal elastomer actuators, are typically dry and driven by electric field or coulomb forces. The following includes several of these polymer actuators that are the most promising for microrobotics.

Overall, polymer microactuators can operate in salt solutions, blood plasma, urine, and cell culture media; can be soft; and can have large strain and energy density at low voltages, which make them attractive for biomedical and soft microrobot applications.

6.3.1 Conductive polymer actuators (CPAs)

Conductive polymers have conductivity in the range of $10^{-12} - 10^{7}$ S/cm, where they have high conductivity when they are doped. Some common conducting polymers are polyaniline, polypyrrole (PPy), and polyacetylene. These materials are semiconductors in their pure form. However, upon oxidation or reduction of the polymer, their conductivity is increased significantly. When electrical current is applied, they can reversibly change their volume due to interactions between polymer chains, conformation of chains, or insertion of counter-ions. The most common CPAs are Au-PPy bilayers, which use the last volume change mechanism. Such bilayers are less than 1 μm thick, and the Au layer acts as the electrode and the structural layer. They change their volume by around 2% upon oxidation ($V > 0$, e.g., $V = 0.35$ V) and reduction ($V < 0$, e.g., $V = -1$ V); counter-ions move outside of the PPy layer during oxidation and inside of the PPy layer during reduction. Such a volume change can induce in-plane (e.g., 0.45-3%) and out-of-plane (e.g., more than 35%) strains depending on their boundary conditions. Their response time depends on polymer layer thickness (diffusion time controls the ion transport).

For electrical actuation of conductive polymer actuators, the typical counter electrode is an Au-coated silicon wafer, the reference electrode is an Ag/AgCl electrode, and the working electrode is the Au layer of the actuator. The electrolyte has a pH more than 3. Jager et al. used such polymer microactutors to build a microgripper arm and foldable microstructures [215]. Such a microgripper could lift a mass 180° back and forth at 0.5-Hz bandwidth. Such a conductive polymer microactuator had 0.2% energy efficiency and a limited lifetime (1,000s of cycles) due to Au-PPy layer delamination (Ti-PPy was better by roughening the Au surface for better adhesion).

6.3.2 Ionic polymer-metal composite (IPMC) actuators

IPMCs bend in response to an electrical activation as a result of the mobility of cations in the polymer network. They are typically composed of an ion exchange polymer membrane, such as Nafion™ from Dupont or Flemion™

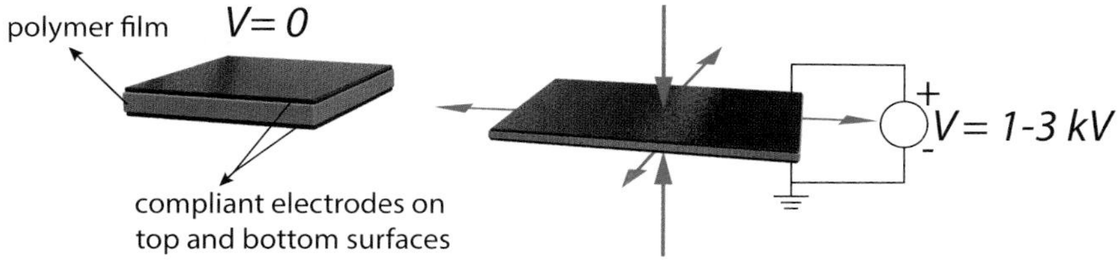

Figure 6.7

Dielectric elastomer actuator contracts vertically and expands laterally when high voltage is applied at its compliant electrodes.

from Asahi Glass, and Pt or Au electrodes deposited on the top and bottom sides of the membrane. When the electric field is applied, the cations move toward the cathode of the polymer membrane dragging molecules of solvents. This motion causes a volume increase in the cathode side and volume reduction in the opposite side. The total motion of the ions and molecules of the solvent creates volume change and motion of the IPMC. They provide large deformation with low applied voltage (1-10 V) at frequencies up to 10s of Hz, and operate best in wet or humid environments, but they can be encapsulated to work in dry environments as well. They are easy to handle and can be prepared in large sheets and then cut to desired shapes and sizes by laser micromachining. When perfluorocarboxylic acid film (140 μm thick) with an Au layer (1.45 μm thick) in both sides is actuated in water, it could even bend at a bandwidth higher than 100 Hz with 10 million cycles durability. Many milliscale swimming robots and soft grippers have been actuated using IPMC actuators.

6.3.3 Dielectric elastomer actuators (DEAs)

Dielectric elastomers are a type of EAPs that use an electric field across a rubbery dielectric material with compliant electrodes to contract. When a high voltage is applied at their compliant electrodes (e.g., carbon-impregnated grease), the Maxwells stress is induced on the elastomer due to the electric field pressure from free charges on the surface of insulating materials, which contracts it vertically and expands laterally, as shown in Figure 6.7. They yield high actuation pressure (0.1-2 MPa), fast response time (<1 ms), high efficiency (up to 80%-90%), high strain (30%-250%), high energy density (0.15 J/g), and low density/weight (1,000 kg/m^3).

Typical dielectic elastomers used in DEAs are silicone elastomers (HS3 from Dow Corning and CF19-2186 from NuSil Technology) and VHB 4910 acrylic elastomer from 3M. VHB 4910 showed the best actuation results, with more than 100% strain (117%-215%), 7 MPa pressure, 3 MJ/m^3 energy density, and 30-40 Hz bandwidth [216]. Such an actuator has been used to actuate microvalves, flapping mechanisms, and soft miniature robots. Although DEAs have many advantages, they have some issues to resolve to be used as a microrobot actuator: carbon-impregnated grease electrodes are not robust and high voltages (3-4 kV) are required. Recent DEAs have also used hyperelastic materials and created large strains using material instabilities [217]. An extensive recent review of soft actuators for small-scale robotics covers all possible existing soft actuation methods and performances [218].

6.4 MEMS Microactuators

Since 1990, MEMS technology has enabled many capacitive/electrostatic, thermal, and magnetic microactuators using optical lithography batch microfabrication processes and electroplating. For electromagnetic actuation, micropatterned areas are electroplated with Cu to create microcoils inducing local magnetic fields to actuate a magnetic Ni or Co structure on a given robot or device, which could be sputter coated or electroplated locally.

For thermal actuation, one- or two-layer metallic structures can be bent due to the heat induced by passing an electrical current through them, as shown in Figure 6.8. Same structures could also be heated up remotely using a focused laser beam. In a one-layer metal beam structure, the geometrical asymmetry in two arms of a cantilever beam (Figure 6.8(a)) can create mechanical bending because each section heats up and expands differently due to their significantly different electrical resistance ($R = \rho_e l/A$, where ρ_e is the material conductance, l is the length, A is the cross-sectional area). Thus, the temperature difference and length increase ($\Delta l = \alpha l \Delta T$) of each arm is different, which induces a bending moment on the overall beam structure. In a bi-layer beam structure, each layer needs to have a significantly different coefficient of thermal expansion α. Such two-material combinations are typically Al-Si, two different metals, and polyimide on a metal or another polyimide. In the bimorph design shown in Figure 6.8(b), radius of curvature R_c of the bent beam can be

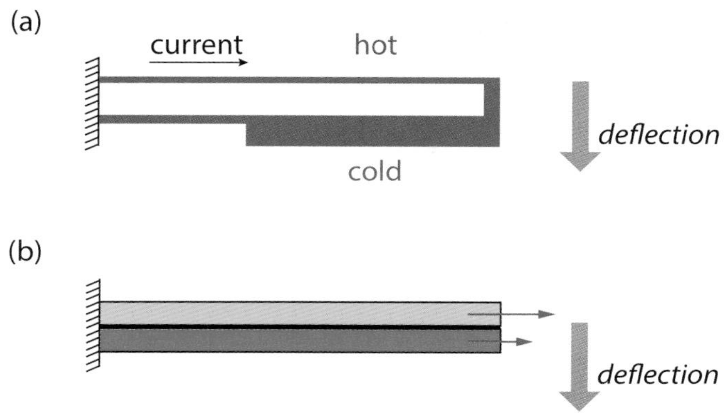

Figure 6.8

Side-view sketches of two thermal microactuator designs: (a) One-layer metal cantilever beam bends when heated because each beam arm heats up and expands differently due to their significantly different geometry and electrical resistance. (b) Bi-layer cantilever beam consists of two layers with significantly different coefficients of thermal expansion.

modeled as

$$R_c = \frac{(t_1 + t_2)^2}{6(\alpha_1 - \alpha_2)t_1 t_2 \Delta T}, \tag{6.30}$$

where t_1 and t_2 and α_1 and α_2 are the layer 1 and 2's thicknesses and coefficient of thermal expansions, respectively. Typical thermal actuators are slow to cool down, and therefore their response is slow. However, scaling them to micron scale makes their surface area to volume ratio much higher and thus their response time much faster.

The most common MEMS microactuator is the electrostatic comb-drive actuator [219]. As we saw in Section 2, electrostatic forces scale favorably for small gaps and actuator sizes. Although they require high voltages, such as 100-150 V, comb-drive actuators can provide high force output and displacement at high bandwidth. They are easy to microfabricate using DRIE type of optical lithography-based microfabrication techniques. Figure 6.9 shows comb-drive actuator design parameters, where the capacitance $C(x)$ and electrostatic force F_{el} can be modeled as

$$C(x) = \frac{2N\varepsilon_0 h(L_0 + x)}{g_0} \tag{6.31}$$

$$F_{el} = \frac{N\varepsilon_0 h(L_0 + x)}{g_0} V^2 = k_x x, \tag{6.32}$$

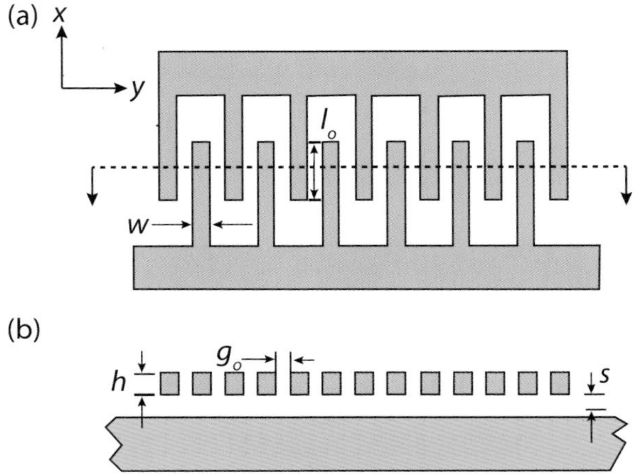

Figure 6.9

Top-view (above) and side-view (bottom) drawings of an electrostatic comb-drive microactuator design with N fingers.

where x and k_x are the comb displacement and stiffness in x-direction, N is the number of fingers, L_0 is the initial overlap length of each finger, h is the finger thickness, g_0 is the gap distance between fingers, and V is the applied voltage, respectively. The x-displacement and F_{el} can be increased linearly by higher N. For stable operation with no y-direction bending, the following condition needs to be held:

$$k_y > \frac{2N\varepsilon_0 h(L_0 + x)}{g_0^3} V^2. \tag{6.33}$$

6.5 Magneto- and Electrorheological Fluid Actuators

Magneto- and electrorheological fluids (MRFs and ERFs), discovered in the late 1940s, are a class of colloidal dispersions which exhibit large reversible changes in their rheological behavior when subjected to a magnetic or an electric field, respectively [220]. Particle-type ERFs consists of micron size highly dielectric particles in insulating silicone oil while homogeneous-type ERFs are made of low molecular or macromolecular liquid crystals. MRFs are mainly dispersions of microparticles made of a soft magnetic material (e.g., carbonyl iron) in a carrier silicone oil. MRFs and ERFs exhibit a pseudo-phase change

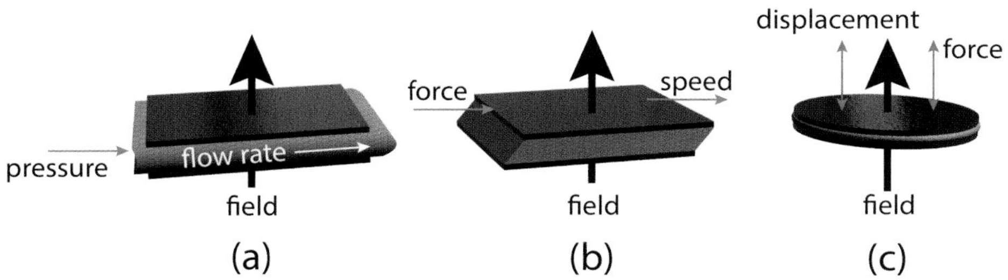

Figure 6.10

Basic operation modes of MRF and ERF actuator designs: (a) flow, (b) shear, and (c) compression modes.

from liquid phase to solid phase by dramatic increase in flow viscosity (up to several magnitudes order) because the initially dispersed microparticles self-assemble into vertical particle chains when a magnetic or electrical field is applied, respectively. Their response time (bandwidth) is fast (1-4 ms), they consume low power, and they create strong shear forces. However, they require high electric (1-4 kV/mm) or high magnetic (0.1-1 T) fields.

MRF or ERF actuators have different modes of operations, as illustrated in Figure 6.10. In the flow mode (Figure 6.10(a)), the fluid between two parallel fixed plates can change its flow rate significantly when the field is applied due to the drastic increase in the fluid viscosity. In the shear mode (Figure 6.10(b)), one of the two parallel plates is movable laterally so that the viscosity change due to the applied field can change the shear force and speed significantly on the movable plate. Finally, in the squeeze (compression) mode (Figure 6.10(c)), the viscosity change can induce force and displacement on the movable place vertically. These actuators have been used in active suspension and damping systems, clutches, valves, brakes, in-pipe crawling miniature robots, and haptic devices.

6.6 Others

Other promising on-board actuation methods that can be scaled down to microscale can be environmental stimulus-based driven actuators, such as paper-based and other composite actuators, driven by the humidity or temperature change in the environment [221].

Table 6.4

Comparison of on-board microactuators, driven electrically (high: • • •; medium: ••; low: •; CD: MEMS comb-drive actuator)

Type	Force	Strain	Speed	Efficiency	Durability	Voltage	Current
Piezo	•••	•	•••	••	••	•••	•
SMA	•••	••	•	•	•	•	•••
CPA	•	•••	•	•	•	•	••
IPMC	•	•••	••	••	••	•	••
DEA	•	•••	•••	•••	••	•••	•
Thermal	•••	••	••	•	•••	•	•••
CD	•••	••	•••	••	•••	•••	•
ERF	•••	•	•••	•••	••	•••	•

6.7 Summary

Design parameters for on-board microactuation methods for a given robotic application are output mechanical force or torque, mechanical power density, output strain or displacement, response time (bandwidth), non-linearity, energy efficiency (i.e., power consumption), durability, operation environment requirements, form factor, weight, and materials (e.g., biocompatibility or biodegradability). After defining the requirements of the given small-scale robotic application related to these design parameters, the proper on-board actuation method with relevant transmission mechanism (if applicable) and load should be selected and then optimized to achieve the requirements with the best performance. Table 6.4 gives a summary comparison of above on-board microactuators for microrobotic applications.

6.8 Homework

1. Section 6.1.2 describes the design and fabrication of PZT-5H (polycrystalline soft piezo material) and PZN-PT (single crystal piezo material) ceramic-based piezoelectric unimorph actuators to design and build a flapping-based flying insect robot. After reading this section, answer the following questions:

a. For a PZT-5H-based unimorph piezoelectric actuator with dimensions of $l = 20$ mm, $w = 4$ mm, $h_p = 127$ μm, and $V = 200$ V, calculate the optimal steel thickness (h_s), DC (low-frequency) tip deflection (δ), blocking force (F_b), resonant frequency (f_r), and bending stiffness K_m using Equation 6.15 and the PZT-5H and steel parameters given in Table 6.2. If the quality factor of this actuator (Q) is measured as 15, then find the actuator tip deflection at resonance (δ_r) for $V = 10$ V.

b. Explain how you can use the same piezoelectric unimorph actuator also as a sensor to detect the aerodynamic forces on the wing. Briefly discuss whether you could detect the low-frequency vibration or force at the wing and why.

c. List the pros and cons of piezoelectric actuators to use them for miniature robot applications in general.

2. We mentioned many on-board actuator types or designs with their pros and cons. For following miniature robotic applications, discuss which actuator designs you would use by explaining the reasons briefly.

a. On a medical ultrasound imaging head.

b. On a millidevice that harvests electrical energy from the vibrations in the environment.

c. On a leg of a walking millirobot on the earth and moon surfaces.

d. On a flapping tail of a fish-like swimming millirobot.

e. On a miniature positioner with nanometer precision.

f. On an undulating body of a milli/microswimmer.

g. On a jelly fish-like milli/microswimmer inside physiological fluids.

h. On a soft crawling millirobot.

i. On an active catheter device inside the vascular system.

j. On a microdevice for GHz frequency RF-relay switching.

7 Actuation Methods for Self-Propelled Microrobots

This chapter includes possible actuation methods that could self-propel a mobile microrobot in a liquid medium. Self-propulsion methods can use self-generated local gradients and fields or biological cells as the actuation source in proper liquid environments. They do not require any on-board electrical power source, electronics, processor, and control circuitry, which make them promising actuation methods for mobile microrobots down to a few microns and even a sub-micron scale.

The dynamics and motion speed of swimming microrobots (microswimmers) with the following self-propulsion methods are determined by the amount of self-generated and viscous drag forces. Therefore, the robot design and operation parameters, such as robot body geometry and size, viscosity of liquid medium, fuel composition, and temperature, need to be optimized for enabling high-speed swimming locomotion for future applications.

7.1 Self-Generated Gradients or Fields-Based Microactuation

Self-generated gradients or fields have been started being used for fluidic propulsion of microscale robots since 2004 [42]. There is a wide range of possible self-generated fields or gradients in liquid media, such as self-electrophoresis, self-diffusiophoresis, self-acoustophoresis, and self-thermophoresis to propel microrobots fluidically.

7.1.1 Self-electrophoretic propulsion

Microrobots with patterned surfaces containing different specific materials (e.g., spherical Janus microparticles and bimetallic microrods) can generate concentration gradients of ions (protons, halide ions, etc.) through bipolar electrochemical reactions at the two ends of the particles or rods (Figure 7.1). The resulting electric field induces motion of the motors through electrophoresis. Such propellers are also called *catalytic micromotors*. For example, bimetallic (e.g., Au-Pt) metal micro/nanorods can propel inside a hydrogen peroxide solution (H_2O_2) using this principle. Electrocatalytic decomposition of H_2O_2 produces hydrogen cations at the anode surface and consumes them at the cathode, yielding an asymmetric ion distribution in the motor vicinity. Such an actuation method always requires fuel, such as hydrogen peroxide.

Electrophoresis describes the transport of micro/nanoscale entities, such as micro/nanoparticles, DNA, and cells, in liquids. In electrophoresis, charged

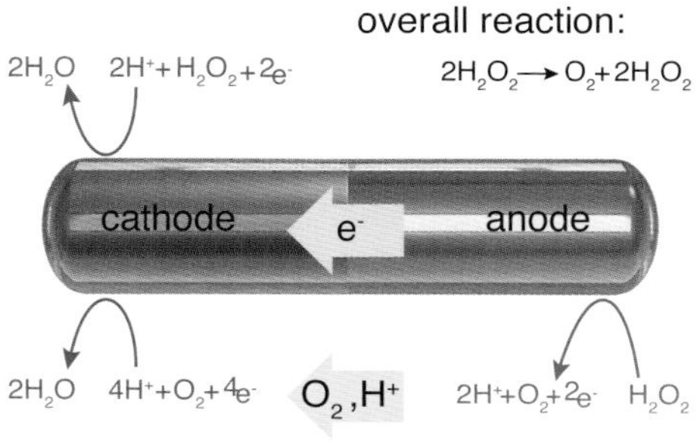

Figure 7.1

Self-electrophoresis, where E is the electric field and H^+ and e^- show the direction of flow of ions and the electric field, respectively.

micro/nanoparticles migrate relative to a fluid due to a spatially uniform electric field (E), and their velocity (v) is governed by the Smoluchowsky equation for particles with thin double layers as [222]

$$v = \frac{\zeta \epsilon}{\mu} E, \tag{7.1}$$

where ζ is the zeta-potential of the particle surface which is related to the surface charge, ϵ is the permittivity of the medium, and μ is the dynamic viscosity of the liquid. In recent decades, many micro/nanoscale systems have exploited the concept of electrophoresis. Unlike ordinary electrophoresis, these self-driven particles do not respond to an externally applied electric field; rather they generate a local electric field through chemical gradients and move in response to this self-generated electric field.

Such an actuation method that catalytically converts chemical energy to fluidic propulsion was first proposed by Paxten et al. [4] in 2004 with Au-Pt nanorods with 2-3 μm length and 300 nm diameter and Fournier-Bidoz et al. [223] in 2005 with Au-Ni nanorods of similar dimension. These nanorods were observed to self-propel in dilute H_2O_2 (a few wt%) with a mean speed of around 10 μm/s. They were the first artificial microsystems to catalytically convert chemical energy to autonomous movement.

In self-electrophoresis, the charged microparticle moves in a self-generated electric field as a result of an asymmetric distribution of ions [222]. For example, in the case of the Au-Pt bimetallic nanomotors, the oxidation of H_2O_2 preferentially occurs at the anode (Pt) end and the reduction of H_2O_2 (and O_2) at the cathode (Au) end. This bipolar electrochemical reaction leads to a higher concentration of protons near the Pt end and a lower concentration near the Au end. Because the protons are positively charged, the asymmetric distribution results in an electric field pointing from the Pt end to the Au end. Therefore, the negatively charged nanorod moves in the electric field, an effect similar to electrophoresis. Although a proton gradient is responsible for the motion of bimetallic motors in H_2O_2 solutions, other ions can also be used to propel motors by the same mechanism. The key is an asymmetric distribution of ions that generates a local electric field.

Bimetallic, self-electrophoretic motors have also been studied in shapes other than rods by fabricating bimetallic spherical Janus particles. The motor speed increases linearly with increasing concentration of H_2O_2. By further breaking the symmetry of the shapes of the motors, it is possible to introduce torque and therefore rotation of the motors. Bi- and trimetallic nanorods can also rotate by evaporating layers of different materials in non-cylindrical geometries.

Fuels other than H_2O_2 have also been shown to propel electrophoretically driven micromotors. Cu-Pt nanorods can move in dilute hydrazine (N_2H_4) and its derivatives, and Cu-Pt and Zn-Pt nanorods or Janus microparticles can move autonomously in dilute I_2 or Br_2 solutions. The lifespans of Cu-Pt and Zn-Pt motors are short due to the corrosion of the active metal segment of the motor. A piece of carbon fiber (7 μm in thickness and 5-10 mm in length) was able to move at the air-water-oxygen interface at 1-10 mm/s for 3 min before stopping [6] in the presence of glucose fuel. Such a terminal glucose-oxidizing microanode and an oxygen-reducing microcathode resulted in a power-generating glucose-oxygen reaction and efficient bioelectrochemical locomotion. Glucose-driven motor systems could be used in biological systems because they would use bio-available fuels: glucose and oxygen. However, their ability to move only at the air-water interface and at high glucose and oxygen concentrations for a short time presents challenges for use in real applications.

The speed of bimetallic nanomotors can be increased up to 150 μm/s by increasing their catalytic surface area [224] or using Ag-Au alloy as the cathode material [225], which could induce larger potential difference between the two electrodes. Moreover, increased temperature due to a heat pulse generated by a laser could increase the motor speed due to faster electrochemical reactions and lower fluid viscosity at elevated temperatures [226].

Many challenges must be solved for this microactuation method for real microrobotic applications. First, most electrophoretically driven motors rely on toxic fuels, such as H_2O_2 or hydrazine. These toxic fuels need to be replaced by biocompatible chemicals, such as glucose. However, significant improvements in efficiency are needed for such motors to be able to move in biological fluids at low glucose and oxygen concentrations. Other possibilities, such as Br_2, I_2, or methanol have been proposed, but glucose remains the most promising one so far. Even if appropriate fuels were found, another major constraint still exists: self-electrophoresis does not work at high ionic strengths. Although these motors can tolerate a low concentration of ionic solutes (e.g., up to 10^{-4} mol/L of Ag(I) ion), electrophoretically driven motors are not compatible for use in biological media with high ionic strength (e.g., around 0.2 mol/L for blood serum). Finally, self-electrophoretically driven motors have an extremely low energy efficiency, on the order of 10^{-8}–10^{-9}. New fuels, motor designs, and propulsion schemes are needed to improve energy efficiency so that these motors can operate at low fuel concentrations.

7.1.2 Self-diffusiophoretic propulsion

Microrobots can be propelled by self-generated chemical concentration gradients, a phenomenon called self-diffusiophoresis. In diffusiophoresis, the motion of particles is driven by a concentration gradient of solutes [227]. Diffusiophoresis has two types, electrolyte and non-electrolyte, in which the molecules contributing to the gradient are charged or uncharged, respectively. Chemical reactions taking place at surfaces consume reactants and generate products, leading to concentration gradients that could propel microrobots, as shown in Figure 7.2. Here, the term self-diffusiophoresis is used because the concentration gradient is generated by the robot interacting with the liquid medium.

Electrolyte diffusiophoresis, which is the more commonly exploited in microrobot propulsion, was first experimentally demonstrated by Ebel et al. [228] in 1988. In such diffusiophoresis, a charged particle is driven by a concentration gradient of ionic species. Because the cations and anions diffuse at

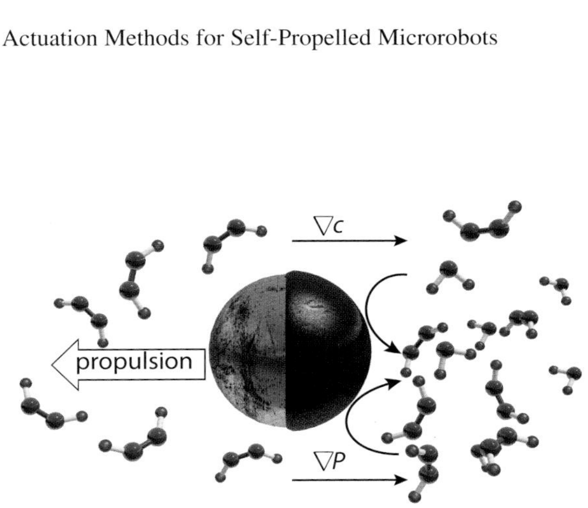

Figure 7.2
Self-diffusiophoresis, ∇C, the concentration gradient, inducing the pressure gradient ∇P to propel the microrobot body.

different rates, an electric field arises that can propel a charged particle. Moreover, cations and anions interact with the double layer of the charged particle differently, resulting in a pressure that moves the particle. In most cases, chemophoretic effects are negligible, and the directions of diffusiophoretic flows are governed by the electrophoretic effect unless the diffusivities of the cations and anions are similar.

Self-diffusiophoretic propulsion is most widely observed for spherical Janus particles propelled by chemical reactions on their surfaces [229]. As examples, from 75-nm to 10-μm diameter particles made of polystyrene (PS), SiO_2, amberlite, and mesoporous silica nanoparticles were half coated with metal (such as Pt, Ag, Au, and Ir) nanofilms to create Janus particles that propelled by self-diffusiophoresis in 0.5-30% H_2O_2, 1M NaOH, or 10^{-7}10% N_2H_4.

Volpe et al. demonstrated light-actuated Janus micromotors in critical mixtures of water and 2,6-lutidine [230]. Janus micromotors were prepared by depositing an Au layer on one of the hemispheres of paramagnetic silica particles, and functionalizing the Au surface with carboxylic thiols to obtain hydrophilic caps. When the Janus micromotors were irradiated by external light (λ = 532 nm), they observed ballistic trajectories of the Janus particles. The swimming mechanism was addressed in a separate study [231], which concluded that the illumination of the Janus motors produces a local asymmetric demixing of a critical binary mixture that generates a concentration gradient around the particle, thereby producing locomotion by a self-diffusiophoretic mechanism.

For self-propulsion applications in biocompatible liquids, a recent approach is self-propelling Janus particles by enzymatic reactions. Polystyrene microparticles with 0.8 μm diameter were functionalized with the enzymes urease and catalase using a biotinstreptavidin linkage procedure [28]. These enzymes were selected due to their robustness and relatively high turnover rates at room temperature. The enzyme-coated particles showed enhanced diffusivity by around 22% in urea solutions due to active propulsion. Furthermore, Ma et al. [232] showed self-propulsion based on Janus hollow mesoporous silica microparticles powered by the biocatalytic decomposition of urea at physiological concentrations. The directional self-propelled motion lasted longer than 10 minutes, with an average velocity of up to five body lengths per second. The speed of the micromotor was controlled by chemically inhibiting and reactivating the enzymatic activity of urease. These example propulsion methods enabled by a chemophoretic mechanism generated by biocatalytic reactions on the surface could be a potential solution for operation inside physiological fluids for future biomedical applications.

7.1.3 Self-generated microbubbles-based propulsion

Gas microbubbles self-generated asymmetrically on the microrobot surface through chemical catalysis or other means can propel microrobots fast and efficiently. In the case of microbubbles self-generated by chemical catalysis, self-propulsion arises from the recoil of gas bubbles produced asymmetrically on the particle surface through chemical catalysis [14, 229] (see Figure 1.5(a)). Bubble-propelled micromotors can be adapted to numerous materials (SiO_2-Pt, Mg-Pt, Mg-Au, and Al-Pd) and shapes (rolled-up microtubes, Janus microspheres, and microtubes) fabricated through electrodeposition with lengths ranging from 5 to 50 μm. Ismagilov et al. [233] moved milliscale objects by using Pt-covered porous glass segments on PDMS surfaces in H_2O_2, and attributed the movement of their object to the recoil force of the O_2 bubbles released from the Pt catalyst. A similar mechanism was also observed for Ni-Au nanowires and carbon microtubes that are modified selectively at one end with a Pt cluster in H_2O_2 [234]. Moreover, the motion of redox Janus particles driven by Al-water [235] or Mg-water reactions [236] follows a bubble propulsion mechanism, as evident by the observable generation and detachment of hydrogen (H_2) bubbles.

The microtubular robots contain active materials, such as Pt on their inner hollow walls for the decomposition of chemicals into gas molecules. Those materials can be catalytic or non-catalytic and decompose diverse fuels (mainly H_2O_2). Such bubble propulsion occurs in three stages. First, the fuel solution interacts with the catalytic material and O_2 accumulates and grows as microbubbles. Next, the bubbles migrate toward and are released from the open end, inducing a finite distance motion step. Motion velocity is linear to the product of the bubble radius and frequency.

Partially coated Al-Ga binary alloy microspheres prepared via microcontact mixing of aluminum microparticles and liquid gallium can self-propel in water towards biomedical applications in physiological fluids [237]. The ejection of hydrogen bubbles from the exposed Al-Ga alloy hemisphere side, upon its contact with water, can provide a powerful directional propulsion thrust. Such spontaneous generation of hydrogen bubbles reflects the rapid reaction between the aluminum alloy and water. The resulting water-driven spherical motors can move at speeds of 3 mm/s (i.e., 150 body length per second) while exerting large forces exceeding 500 pN.

7.1.4 Self-acoustophoretic propulsion

Self-acoustophoretic motors are propelled by asymmetric steady streaming of the fluid around them in an acoustic field; thus, they can operate in water and biological fluids towards biomedical applications. Using low-power ultrasonic acoustic waves, microrobots can experience acoustic radiation forces that move them to the pressure nodes (or antinodes) as a result of pressure gradients. When the acoustic excitation meets the criteria to form standing waves, the radiation force is the strongest. Wang et al. proposed a self-acoustophoretically propelled microrobot in 2012 using around 4-MHz frequency ultrasound waves [20]. In this system, asymmetrically shaped Ru-Au metallic microrods (a few microns long) were suspended in water in an acoustic chamber. Vertical standing waves levitated the rods to a plane where the pressure was minimum. In that plane, the microrods had an axial motion at speeds up to 200 μm/s in water. They formed specific patterns in the nodal plane as a result of nodes and anti-nodes within the plane. The composition of the microrods was found to significantly affect their movement, with only metal microparticles showing fast axial motion. The back end of the microrods was concave shaped while the front end was convex. It was believed that both ends scattered the acoustic waves differently, leading to a pressure gradient that is high at the concave end and low at the convex end, with a difference of around 1 Pa. Such a pressure gradient is believed to create the strong axial propulsion of the metallic rods.

7.1.5 Self-thermophoretic propulsion

Temperature gradients can also be used by microrobots to create motion, a mechanism referred to as self-thermophoresis. Thermophoresis, or the Soret effect, has been known and studied for more than 150 years. However, the study of thermophoresis has been mostly carried out in macroscopic systems in which colloidal particles collectively migrate in an externally established thermal gradient. Various microswimmers are propelled by self-generated temperature gradients in water [17, 238, 239]. Jiang et al. reported the propulsion of Au-silica Janus microspheres by self-thermophoresis under laser irradiation (defocused laser beam at 1,064 nm) [17]. The Au cap absorbed the laser irradiation and generated a local thermal gradient of temperature (around 2 K across the particle) that induced motion. Qian et al. theoretically modeled the temperature gradient of a Janus particle irradiated with a laser and illustrated 3D motion of an Au-PS Janus particle experimentally [239]. Next, Baraban et al. used an AC magnetic field to heat Permalloy-capped silica particles in solution and observed propulsion [238]. The temperature difference across the particle was estimated to be 1.7 K. In these demonstrations, the microparticles moved away from their heated side, exhibiting a positive thermophoresis effect. Moreover, Golestanian studied the collective behavior of self-thermophoretic microswimmers using a stochastic formulation [240]. He found that thermorepulsive swimmers could organize into different structures while thermoattractive swimmers became unstable.

7.1.6 Self-generated Marangoni flows-based propulsion

The Marangoni effect is the mass transfer along an interface between two fluids due to surface tension gradient. Marangoni flows are those driven by such surface tension gradients. In general, surface tension σ depends on both the temperature and chemical composition at the interface. Therefore, Marangoni flows may be generated by gradients in either temperature or chemical concentration at an interface. In the case of temperature dependence, this phenomenon may be called thermocapillary convection. In the case of chemical composition-based surface tension gradient, surfactants are typically used. Surfactants are molecules that have an affinity for interfaces; common examples include soap and oil. Owing to their molecular structure (typically a hydrophylic head and hydrophobic tail), they find it energetically favourable to reside at the free surface. Their presence reduces the surface tension. Therefore, gradients in surfactant concentration result in surface tension gradients. Thus, surfactants generate a special class of Marangoni flows. For example, a soap boat can be simply generated by coating one end of a toothpick with soap, which acts to reduce surface tension. Due to non-homogeneous surface tension around the boat, a net lateral propulsive force is induced by driving the boat away from the soap. A similar Marangoni propulsion can arise in nature also: certain water-walking insects eject surfactant and use the resulting surface tension gradients for rapid water surface propulsion. Moreover, when a pine needle falls into a pond, it is propelled across the surface similarly due to the influence of the resin at its base, decreasing the local surface tension.

Consider a flat floating robot body with perimeter C in contact with the free surface. For σ as a force per length in a direction tangent to the surface, the total tangential surface tension force acting on the body is [241]:

$$F_c = \int_C \sigma \mathbf{s} dl, \tag{7.2}$$

where $\mathbf{s}$ is the unit vector tangent to the free surface and normal to C, and dl is an incremental arc length along C. If σ is constant, then this line integral vanishes by the Divergence Theorem. However, if $\sigma = \sigma(x)$, then this equation can result in a net Marangoni propulsive force, as in the case of the soap boat.

Many milli/microscale floating robots have been propelled in the water-air interface using the surfactant-based Marangoni effect [242–246]. As an example, a pipette tip as a milliswimmer was demonstrated based on rapid depolymerization of polymers that could generate Marangoni flows [242]. The poly(2-ethyl cyanoacrylate) (PECA) was synthesized by isocyanate ions generation. The FDA-approved PECA is known as a biocompatible and biodegradable polymer that is non-toxic and non-immunogenic to the human body. The primary surface active product (ethanol) released from the open end of the pipette tip powered its movement. Thus, the surface tension force was caused by ethanol among the products that lowered the surface tension and broke the symmetry to induce motion at the air-water interface.

Surfactant-based Marangoni propulsion is highly efficient and fast. Its main drawback is its limited actuation duration due to the limited volume of the surfactant source that can be stored on the miniature robot. Moreover, such a robot always moves and requires additional mechanisms to stop and steer it.

Marangoni propulsion can also be used to propel microswimmers under water if microbubbles were attached to the microrobot body and have local surface tension gradients induced by local surfactant, heating, or temperature gradient.

7.1.7 Others

Linear motion of conductive Janus spheres can be induced by bipolar electrochemistry (BPE) [247]. The fundamental idea of BPE is based on the fact that when a conducting object is placed in an aqueous solution in which an external electric field is applied between two electrodes, a maximum polarization voltage ΔV occurs between the two hemispheres of the object. The value of $\Delta V = El$, where E is the electric field and l is the characteristic dimension of the object. When the ΔV value is appropriate, redox reactions occur at the opposite poles of the particle. For example, asymmetric bubble propulsion as a result of water electrolysis can result in translational motion [247]. Additionally, the bubble propulsion speed can be enhanced by adding to the solution sacrificial compounds, which are easier to oxidize or to reduce than H_2O molecules, and leads to bubble formation in only one of the hemispheres. Moreover, the propulsion of Pt-PS dimers was demonstrated under an electric field in 2014 [248].

7.2 Bio-Hybrid Cell-Based Microactuation

Cells, such as contractile cardiac, smooth and skeletal muscle cells, swimming or surface crawling microorganisms, such as bacteria and algae, and some other motile cells can be attached to synthetic microrobot bodies and actuate them using the chemical energy (adenosine triphosphate, ATP) inside their cell body or in the environment. Such an actuation method is called *bio-hybrid* (biotic and abiotic materials are integrated to each other) cell-based microactuation [52]. Biological cells have evolved over millions of years to have highly efficient, robust, and agile locomotion at the microscale. The energy conversion efficiency of biological motors is orders of magnitude greater than that of existing synthetic (non-biological) micromotors. Biological cells can be used to convert chemical energy into mechanical work and can respond to forces, mechanical strain, and chemicals in their environment through integrated sensing and control pathways. The goal of bio-hybrid microactuation design is to utilize these cellular functionalities in a microrobot to create new bio-hybrid microrobots, such as microgrippers, microrotors, micropumps, microswimmers, and microwalkers, which could work in physiological or complex real-world environments..

The bio-hybrid actuation approach has numerous advantages. One of the significant benefits of incorporating live cells into a microrobot is that they enable an untethered operation. Biological cells are driven by chemical energy. As long as the proper nutrients are supplied in the environment, they can convert chemical energy into mechanical work. Furthermore, biological cells have the ability to functionally adapt to their environment, are capable of self-repair and self-assembly, and have developed sophisticated sensing and actuation mechanisms, which cannot be artificially replicated with currently available technologies. By incorporating living cells into a microrobot, we can take advantage of their superior actuation and sensing capabilities.

The development of a bio-hybrid microrobot requires two main physical components: (1) live biological cells that can serve as actuators and/or sensors, and (2) an artificial substrate that can provide structural support and aid in the functionality of the robot. Examples of these components are listed in Figure 7.3. In addition to these physical components, a method of control is required to perform tasks, such as navigation and manipulation. Some common control methods include the use of magnetic fields, electric fields, optical stimuli, and chemical stimuli. Here, we discuss these components and show how they can be integrated to form a functional tool or an actuating component in a larger system.

7.2.1 Biological cells as actuators

Not all biological cells are suitable microactuators for a bio-hybrid robot. First, the cells must be capable of generating a force. Two types of cells that can produce such a force are motile microorganisms and muscle cells. Motile microorganisms are driven by biomolecular motors, producing a propulsive force that allows them to swim through fluidic environments or glide on surfaces. They are commonly used as bio-hybrid microtransporters. Muscle cells contract in response to an electric stimulus and have been commonly applied to actuate microstructures, such as micropillars, cantilevers, and thin films. To be a viable candidate for a bio-hybrid robot, the biological cell should be easily cultured and maintained over a wide range of environmental conditions. The

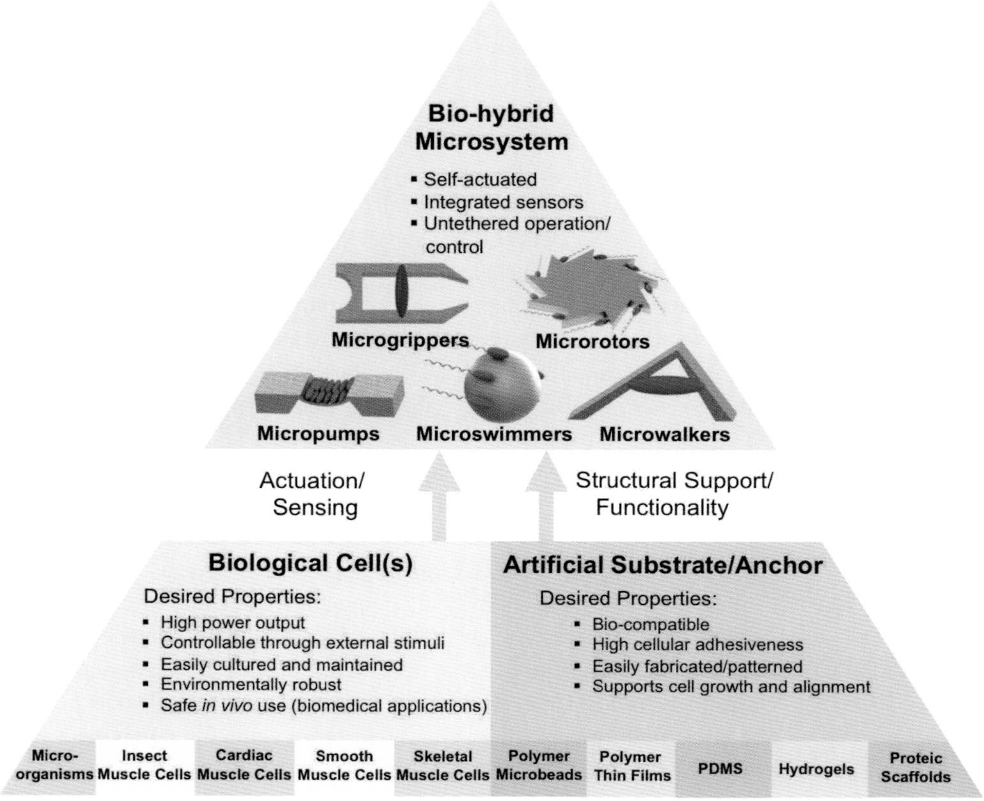

Figure 7.3

Main components of a bio-hybrid microrobot [52]. Copyright © 2014 by John Wiley Sons, Inc. Reprinted by permission of John Wiley & Sons, Inc. These microrobotic systems integrate biological cells with artificial substrates to provide actuation and sensing functionalities.

cells should also generate an actuating force that is repeatable and controllable. The choice of an appropriate biological cell actuator is highly dependent on the intended application of the robot. For example, drug delivery applications require cells that are small enough to travel through the capillaries of the human body and are safe enough for in vivo use, whereas a microgripper for manipulation applications may require cells that produce a large contractile force that is remotely controllable and repeatable. Here, we outline some of the main characteristics of commonly used biological cell actuators and describe the main advantages and disadvantages of each.

7.2.1.1 Microorganism-based actuators

A wide range of microorganisms has been applied in bio-hybrid microsystems [52]. These include flagellated bacterial species, such as *Escherichia coli* (*E. coli*), *Serratia marcescens* (*S. marcescens*), *Salmonella typhimurium*, *Vibio alginolyticus*, and *Bacillus subtilis*; strains of magnetotacic bacteria [249]; and gliding bacterial species, such as *Mycoplasma mobile*. In addition to bacterial cells; strains of protozoa, such as *Paramecium caudatum*, *Vorticella convallaria*, and *Tetrahymena pyriformis*; and strains of algae, such as *Chlamydomonas reinhardtii*, have been applied in bio-hybrid devices. These microorganisms generate forces through the movement of flagella or cilia, which are long appendages that extend from their cell body. Flagellated bacteria rotate their flagella to produce thrust forces ranging from around 0.5 pN for *E. coli* and *S. marcescens* to about 4 pN forces for *Magnetospirillum marine coccus* (MC-1), enabling the cells to swim at speeds as high as 100 times their body length per second. Most of these bacterial species are only 1 to 3 μm in size and therefore, are ideal for drug delivery applications because they can pass through human capillaries, which are typically 4 to 8 μm in diameter. Some of the larger microorganisms, such as *Paramecium caudatum*, can grow as big as 300 μm and can generate a higher thrust force of 27 nN.

The major advantages of using microorganisms as actuators are that they are generally small, are easily acquired through a number of established cell lines, and can survive in a wide range of environmental conditions. Some microorganisms can live in environments with temperatures greater than 100°C, and others can grow in pH values as low as 2 or as high as 11.5. Microorganisms also only require simple nutrients (e.g., glucose) for growth. Furthermore, most of these cells exhibit a taxis response, which is a movement in response to an environmental stimulus. This taxis behavior can be applied as a means to steer the microorganism-driven device. Some common types of taxis responses include chemotaxis (chemicals), magnetotaxis (magnetic field), galvanotaxis (electric field), phototaxis (light), thermotaxis (temperature), and aerotaxis (oxygen), enabling a wide range of steering control methods [52]. A major disadvantage of using microorganisms as actuators is that some strains are pathogenic and can elicit an immune response when used in the human body. This has limited their usage to ex vivo applications so far. However, through genetic modification, non-pathogenic forms of these strains can often be engineered. Furthermore, nine out of every ten cells found in the human body are microorganisms (i.e., human microbiome). For in vivo applications in locations of the body where these cells are naturally found, a wide variety of cells could potentially be applied towards these bio-hybrid robots.

7.2.1.2 Muscle-based actuators

Muscle cells (also called myocytes) produce a contractile force, ranging from a magnitude of a few micronewtons for a single cell up to several hundred micronewtons for engineered muscle tissue [52]. They contract at frequencies of 1 to 5 Hz and exhibit a prolonged contraction state called tetanus when stimulated at higher frequencies. Three main types of muscle cells are found in vertebrates: cardiac, smooth, and skeletal. These cells are around 20 μm in diameter and can range in length from 100 μm to a few millimeters. Although individual muscle cells have been used to drive simple actuators, the cells are more commonly applied in the form of cell sheets or as 3D muscle tissue constructs. The thickness of these constructs is limited to less than a millimeter due to a lack of a vascular system in engineered tissues. Because the generated power is proportional to the cross-sectional area of the tissue construct, much research has been invested into techniques for incorporating an engineered vascular system so thicker tissues can be produced.

Of the three types of muscle cells, cardiac and smooth muscle were first explored as bio-hybrid actuators because these cells contract spontaneously and do not require an external driving stimulus. However, as control methods advanced, spontaneous contraction was seen as more of a disadvantage limiting their controllability. Localized control of these cells within a tissue construct is also limited because these cells are coupled via gap junctions, which causes an electric impulse to propagate between cells, leading to synchronous stimulation. Skeletal muscle cells, in contrast, do not contract spontaneously and can be regulated using an electric potential. In the human body, skeletal muscle tissue is comprised of motor units, each of which can be stimulated individually, allowing modulated control of contractions. This level of control has not yet been achieved in engineered bio-hybrid systems; however, localized stimulation of engineered muscle tissue constructs is possible, e.g., by using an underlying microelectrode array.

One of the major disadvantages of using mammalian muscle cells for bio-hybrid applications is that the culture environment has to be tightly controlled.

Table 7.1

Main advantages and disadvantages of common control methods for bio-hybrid microactuators [52]

Methods	Advantages	Disadvantages
Magnetic	Fast response time ($\leq$ 1 s)	Selective, parallel stimulation difficult
	Remotely controllable, non-invasive	Heat management of electromagnets
	Penetrates most environments	Requires modification of non-magnetic cells/substrates
	Environmentally robust	Requires external control system
Electrical	Fast response time ($\leq$ 1 s)	Possible generation of harmful chemicals/gases/heat
	Selective, parallel stimulation possible	Invasive (requires (close) contact of electrodes)
	Remotely controllable	Requires external control system
Optical	Fast response time ($\leq$ 1-10 s)	Cannot penetrate all environments
	Selective, parallel stimulation possible	Requires genetic modification of some cells
	Remotely controllable, non-invasive	Requires external control system
Chemical	No external control system required	Selective, parallel stimulation difficult
		Limited steering control
		Response time dependent on chemical diffusion rate
		Requires chemical sensing/response ability

To maintain the viability of the cells, the culture medium must be kept at 37°C with a pH of 7.4 and a 5% carbon dioxide supply, and it has to be replaced every few days to remove byproducts and supply the cells with sufficient nutrients. This maintenance is a major drawback to using mammalian muscle cells in bio-hybrid applications. As a viable alternative, insect muscle cells have recently been proposed for use in a bio-hybrid actuator. Insect muscle cells are more environmentally robust. They have been observed to contract for more than a month without changing the culture medium and can operate over a temperature range of 5 to 40°C, although the contraction frequency was shown to vary with temperature. The cells can also be controlled with an electrical stimulus. Insect muscle cells have been successfully demonstrated as actuators in bio-hybrid devices; however, widespread use of these cells requires the establishment of cell lines and improved culturing techniques. Currently, these muscle cells can only be used as tissue explants because a method to dissociate the tissue has yet to be established.

7.2.2 Integration of cells with artificial components

The performance of a bio-hybrid microsystem is highly dependent on the proper coupling of the cells to the substrate. Adhesion is dependent on a number of factors, including the surface chemistry, surface topography, surface charge, and hydrophobicity. Some types of synthetic materials must be modified to promote cell adhesion. For example, cells attach poorly to hydrogels because the material is hydrophilic. Hydrogels can be modified with RGD (arginine-glycine-aspartic acid) cell adhesion peptides to enhance attachment to cardiomyocytes. Collagen molecules can be chemically attached to PEGDA hydrogels to promote attachment to cardiomyocytes. After surface modification, several methods can be applied for the actual attachment of cells to the artificial substrate. Muscle cells are typically cultured directly onto the substrate material. Thermoresponsive polymers, such as poly(N-isopropylacrylamide) (PIPAAm), can also be utilized to transfer cultured cells to another substrate. PIPAAm transitions from a hydrophobic to a hydrophilic material when the temperature is decreased below 32°C, enabling the controlled release of cultured layers of cells. In contrast to the direct culturing of cells onto an artificial substrate, motile microorganisms are typically attached to objects through stochastic interactions. For example, a blotting technique is often applied to attach bacterial cells to microbeads [3, 70]. In this technique, the microbeads are placed directly onto a plate of swarming cells, and the cells randomly collide with the beads and attach to their surface.

In addition, the maintenance of biological cells is a critical factor affecting the operation lifetime of a bio-hybrid robot. For biological cells to remain viable, they must be fully immersed in culture media. This requires the cells to be isolated and encased if not used in a fluidic environment. For example, an atmospher-operable bio-hybrid gripper can be developed by packaging the actuating insect muscle cells within a sealed capsule that contained 40 μL of culture medium. Mammalian cells also require frequent exchange of culture media and are highly vulnerable to infection if not maintained in a sterile environment. For the successful implementation of bio-hybrid robots, nutrient transport systems must be developed, or biological cells that are more environmentally robust, such as bacterial cells and insect cells, must be used.

7.2.3 Control methods

The modulation and regulation of the actuation force of bio-hybrid microsystems are critical to effectively employ these bio-hybrid robots. Control inputs can be received from an operator remotely or from stimuli in the local environment. Cell-based control methods utilize the sensory pathways of biological

Table 7.2
Average swimming speeds and cargo-to-cell size ratios of cell-actuated microswimmers (blps: body length per second; PS: polystyrene) [52]

Cell Type	Steering Control	Cargo	Mean Speed (blps)	Cargo-to-Cell Size Ratio
C. reinhardtii	Optical	PS microbead	7.2	0.25
E. coli	None	PS microbead	7.2	0.25
Magnetotactic ovoid	Magnetic (0.2 mT)	PS microbead	5.3	1.1
V. alginolyticus	None	PS microbead	2.7	1.5
S. marcescens	None	PS microbead	2.6	5
S. typhimurium	None	PS microbead	2.1	3
Magnetotactic MC-1	Magnetic (0.35 mT)	PS microbead	1.9	1.5
S. marcescens	None	PS microbead	0.4	5
S. marcescens	Chemical	PS microbead	0.7	2.5
S. marcescens	pH	PS microbead	2.0	3.0
V. alginolyticus	None	Liposome	0.3	6.5
Bovine sperm	Magnetic (22 mT)	Ti/Fe microtube	0.2	5
S. marcescens	Electric (8 V/cm)	SU-8 microstructure	0.1	20

cells to elicit a response. Non-cell-based methods provide an external stimulus that induces a response without utilizing the sensing mechanisms of living cells. For example, the directed motion of a device through electrophoresis is a non-cell-based control method. The response time also varies for different control strategies. Some control methods elicit an immediate response, such as the contraction of a muscle cell from an electric potential, whereas other methods result in a delayed response. For example, the chemotactic steering of a bio-hybrid microswimmer occurs over a time scale of tens of seconds. Table 7.1 lists common methods of control along with their major advantages and disadvantages.

7.2.4 Case study: Bacteria-driven microswimmers

Motile bacteria can be attached to the surface of artificial particles or other materials to propel them in stagnant fluids as a microactuator and microsensor. Such bio-hybrid microswimmers offer a unique advantage over non-biological synthetic microswimmers in that the bacterial cells can detect and respond to various types of environmental stimuli. As a result, these bio-hybrid microswimmers can be steered using both environmental stimuli, such as biochemical signals released from a diseased tissue, and external control inputs,

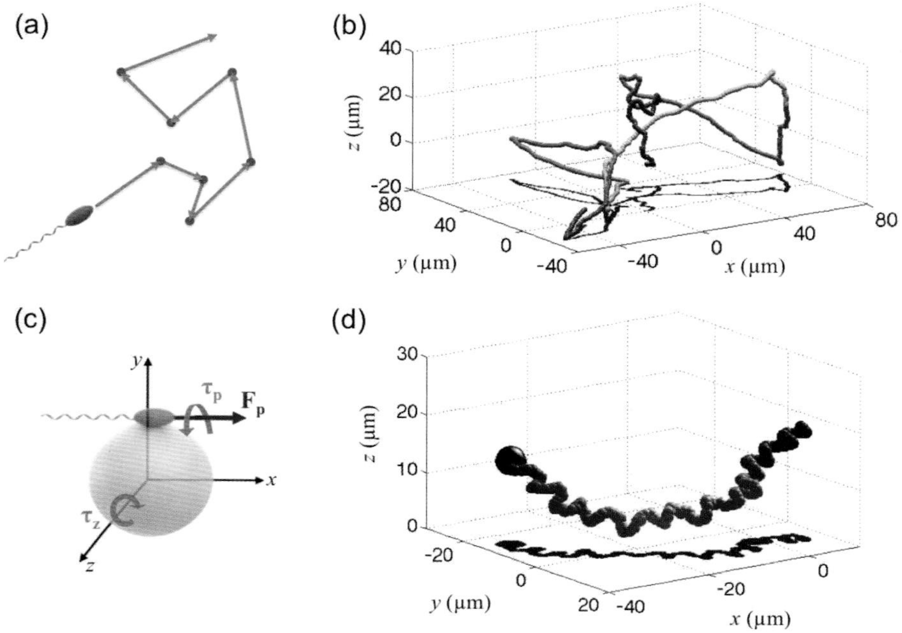

Figure 7.4

Stochastic swimming motion behavior of *S. marcescens* and a *S. marcescens*-propelled microswimmer [52]. Copyright © 2014 by John Wiley Sons, Inc. Reprinted by permission of John Wiley & Sons, Inc. (a) A free-swimming bacterial cell alternates between run states, where it travels in a straight line (red arrows), and tumble states, where it tumbles and reorients in 3D space (blue dots). (b) Experimentally measured 3D swimming trajectory of a free-swimming bacterial cell (*S. marcescens*). (c) Propulsive forces and torques generated by a single bacterium attached to a microbead. (d) Representative 3D helical trajectory of a *S. marcescens*-propelled microbead, which was obtained experimentally [53].

such as magnetic fields and light. Here, we discuss the important considerations in the design of these devices, and propose how the performance of these devices can be optimized based on our current understanding of these systems.

Bacteria as the actuator: Several bacterial strains have been applied in microswimmer applications as listed in Table 7.2. To analyze the design of these devices, we will consider *S. marcescens* as a model bacterium because it is the most commonly used bacterial species in these applications. This bacterium is naturally pathogenic; however, with genetic modification, it can be applied in biomedical applications. Its morphology, motility, and tactic behavior are very similar to *Escherichia coli* (*E. coli*), which is the most understood and documented bacterium type [250]. The bacterium has a rod-shaped cell body 0.7 to 2 μm long and 0.5 to 1 μm in diameter and has 1 to 10 flagella, which project from its cell body in all directions. In the swimming state, the flagella form a single bundle. A propulsive force is generated by the rotation of the flagella, which drives the bacterium forward. The mean free-swimming speed of *S. marcescens* is 26 μm/s, and it can produce a thrust force of about 0.5 pN [53]. The bacterium swims by alternating between run and tumble states, resulting in a random walk, as shown in Figure 7.4(b). In the run state, it moves in a straight line, and in the tumble state, its cell body reorients in space as depicted in Figure 7.4(a). It alters between these states by changing the rotation direction of the flagellar motor. The average tumble rate of *S. marcescens* is 1.34 ± 0.16 tumbles/s, but it can decrease its tumble rate if it is moving in a favorable direction (e.g., food source), resulting in a biased random walk.

Artificial substrate: Several materials can be used as artificial substrates for bio-hybrid microswimmers. Polystyrene microbeads are the most common ones as shown in Table 7.2 because they are easy to model theoretically, commercially available, and can be functionalized with a surface coating or can be fluorescently stained easily. In addition to PS microbeads, microstructures made of SU-8, PDMS, poly(ethylene glycol) (PEG), hydrogel, etc. can be used as the artificial substrate.

Integration of the bacteria with microparticles: The bio-hybrid microswimmer is fabricated by adhering *S. marcescens* to a synthetic microobject. In contrast to bacterial species, such as *E. coli* [251], *S. marcescens* adheres readily to most surfaces through hydrophobic interactions and electrostatic forces [53]. Although surface modification is usually not necessary for *S. marcescens*, it has been applied to enhance the adhesion to other bacterial species or to modify surfaces that are hydrophilic, such as that of hydrogels and liposomes, which are highly non-adhesive. Two chemical modification approaches have been widely used in bio-hybrid microswimmers to facilitate bacteria adhesion. The first method uses a nonspecific binding agent, such as poly-lysine, to enhance binding. The second method applies specific binding, such as through antibody binding or utilizing the strong binding affinity between molecules, such as biotin and streptavidin. Patterning methods can also be applied to isolate bacterial attachment to one portion of a microobject for enhanced propulsion efficiency by not canceling out the propulsion force of the attached bacteria. For example, spherical Janus microparticles are shown to create more efficient propulsion for bacteria-driven microswimmers, where

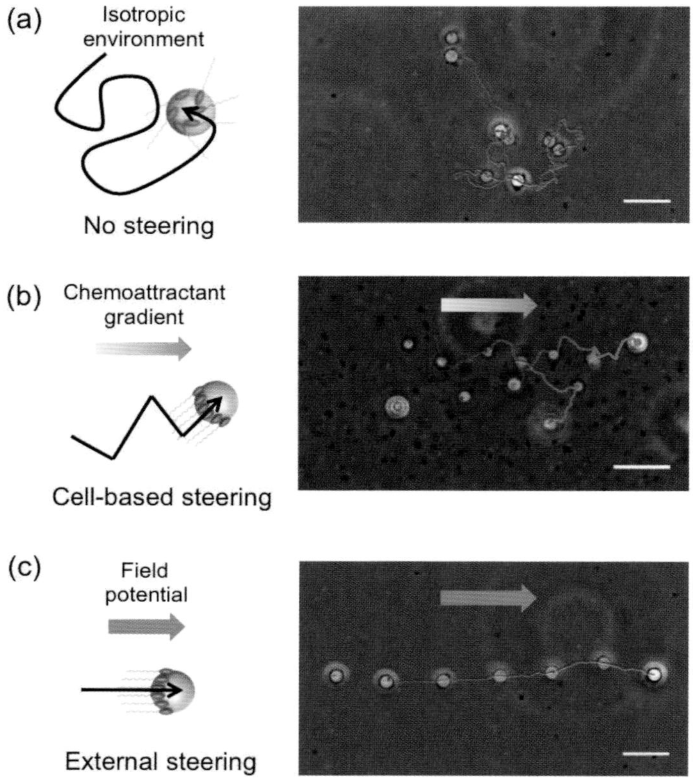

Figure 7.5

Control strategies for a bacteria-propelled microswimmer [52]. Copyright © 2014 by John Wiley Sons, Inc. Reprinted by permission of John Wiley & Sons, Inc. Experimental time-lapse images are shown on the right for microswimmers in (a) an isotropic environment resulting in stochastic motion; (b) a linear chemoattractant (L-aspartate) gradient, which is sensed by the bacteria, leading to a biased random walk; and (c) an applied uniform magnetic field of 10 mT resulting in directed motion. The bacteria are attached to a superparamagnetic microbead in (c) and to polystyrene microbeads in (a) and (b). The red and blue lines indicate the trajectories of the microswimmers. Different microswimmer samples are shown for each environmental condition. Scale bars: 20 μm.

PS microbeads were half-patterned with a surface coating that prevented bacteria attachment [251, 252]. The most common method for attaching bacteria to a microobject is through a blotting technique, where a swarm of cells is placed in physical contact with the microobject. Bacteria randomly adhere to the object through stochastic interactions. When attaching to an object, bacterial cells tend to maximize their contact area. For a smooth object, such as a sphere, they will attach to the surface with the sides of their cell body, which maximizes their contact area. Figure 7.4(c) shows a schematic of a bacterial cell attached to a microbead with the side of its cell body.

Motion and control: The highly stochastic motion of a bacteria-propelled microswimmer can be understood by analyzing the stochastic propulsive forces and torques generated by the randomly distributed and oriented bacteria. We consider the case of single *S. marcescens* cell attached to a microsphere, as shown in Figure 7.4(c) for simplicity. We assume that the flagella are bundled and aligned in a single direction. The rotation of the flagella produces a force and torque on the bead. Because the force is not directed toward the bead's center of pressure, it produces an additional torque on the bead, τ_z. Microswimmers operate in a low Reynolds number environment, and the propulsive forces and torques are balanced by the fluidic drag force and torque. These forces and torques produce a helical swimming motion as shown in Figure 7.4(d). This helical motion is observed for beads propelled by a small number of bacteria [53]. A long time correlation in the motion of bacteria-propelled beads can also be observed (i.e., a significant periodic component in the motion of the beads existed over time intervals larger than 25 s), suggesting that the forces and torques are nearly constant. In the case of microswimmers driven by multiple cells, the net force and torque may not be constant, resulting in a stochastic motion of the microswimmer shown in Figure 7.5(a).

A steering control method is required to guide the motion of the microswimmer. As discussed earlier, several types of steering control methods have been applied to bacteria-propelled devices. Using cell-based control methods, such as chemotactic control, the microswimmer can be directed to conduct a biased random walk, as shown in Figure 7.5(b) [253]. pH-taxis could also be used to steer the bacteria-propelled microswimmers where bacteria swim towards a preferred pH level (around 7.2 for *E. coli*) [22]. External control inputs that do not utilize the cellular sensing pathways can also be applied for steering control. Using an electric or magnetic field potential, a bio-hybrid microswimmer can be guided along the direction of the applied field as shown in Figure 7.5(c) [23]. The bacteria-propelled microswimmer in Figure 7.5(c) consists of cells attached to a superparamagnetic bead. Because *S. marcescens* is not naturally magnetotactic, a magnetic substrate is necessary. An applied magnetic field keeps the magnetic moment of the bead aligned with the field, resulting in a high degree of steering control.

Analysis of the stochastic motion data: The stochastic 3D translational and rotational motion of the bacteria-propelled beads is a 3D random walk. The mean velocity V_{mean} of the beads can be measured from the stochastic motion trajectories (i.e., the mean-squared displacement, ΔL^2, data) to quantify their propulsion performance. In general, the mean-squared displacement of microbeads subjected to stochastic translation and rotation in 3D can be described by [50]:

$$\Delta L^2 = 6D\Delta t + \frac{V_{mean}^2 \tau_R^2}{2}\left[\frac{2\Delta t}{\tau_R} + e^{-2\Delta t/\tau_R} - 1\right], \tag{7.3}$$

where $D = k_B T/(6\pi \mu R)$ is the translational diffusion coefficient, k_B is the Boltzman constant, T is the absolute temperature, and $\tau_R = 1/D_R = 8\pi \mu R^3/(k_B T)$ is the rotational diffusion randomization time. For example, for a 5-μm radius bead moving stochastically in water, τ_R is around 11 minutes.

In short durations when the motion time is much less than the randomization time ($\Delta t \ll \tau_R$), we can express the exponential term by its Taylor approximation in the vicinity of 0, and Eq. [7.3] simplifies to:

$$\Delta L^2 \approx 6D\Delta t + V_{mean}^2 \Delta t^2. \tag{7.4}$$

Furthermore, given that the effect of bacteria propulsion is much greater than the translational diffusion, this equation can be further simplified to:

$$\Delta L^2 \approx V_{mean}^2 \Delta t^2. \tag{7.5}$$

This means, for $\Delta t \ll \tau_R$, any ΔL^2 data of the bacteria-propelled beads plotted against Δt can be fitted to a *quadratic* function where the coefficient gives V_{mean}^2. This is how V_{mean} is computed or measured in the bactaria-driven and other microswimmers with random walk-based stochastic motion at short durations. In such short durations, the beads would have a *ballistic* Brownian motion where the active bacteria propulsion would dominate their translational diffusion.

For long durations ($\Delta t \gg \tau_R$), Eq. [7.3] simplifies to:

$$\Delta L^2 \approx (6D + V_{mean}^2 \tau_R)\Delta t = 6D_{eff}\Delta t, \tag{7.6}$$

where D_{eff} is the effective translational diffusion coefficient of the bacteria-propelled beads. Thus, the beads would effectively have a *diffusive* Brownian motion in long durations when $\Delta t \gg \tau_R$ with a much larger diffusion constant of D_{eff} due to the active bacteria propulsion. In this model, the active particle is assumed to have a negligible active mean rotational speed, ω. However, bacteria attached to the bead surface could induce non-negligible ω. In this case, similar to the translational diffusion coefficient, we would expect a combination $\tau_R^{-1} + \omega^2 \tau_R$ to serve as the renormalized rotational diffusion coefficient.

Stochastic motion modeling of bacteria-propelled beads: In the case of microswimmers driven by multiple cells, as shown in Figure 7.6, the net force and torque are random and not constant, resulting in a stochastic motion of the microswimmer shown in Figure 7.5(a). To model such behavior, let us first model the single bacterium behavior. Single bacterium moves in two distinct states: run and tumble. The run state is characterized by a bacterium bundling its multiple flagella together and rotating them counterclockwise to achieve propulsion in the forward direction, while during the tumble state one or more flagellar motor rotate clockwise and the flagella unbundle, ceasing to generate any propulsive force and thereby allowing momentary Brownian diffusion only. These two operating states can be modeled as a continuous time Markov chain, where state transitions occur in a Poisson manner. Hence, the transition probabilities between states, k_r and k_t, can also be viewed as the arrival rates of the next state, explicitly k_r arrival rate of a run state when the bacteria is tumbling, k_t arrival rate of a tumble state when running. In the absence of chemical stimuli in the medium, the run and tumble phases occur stochastically and, in combination with diffusion, result in the bacteria and bacteria-propelled objects performing a *random walk*.

The inter-arrival times between run and tumble events are exponentially distributed in a Poisson process, and they can be described in the following manner:

$$f(t, \lambda) = \lambda e^{-\lambda t}, \tag{7.7}$$

where t is the time of an arrival event and k is the arrival rate of events. For the two-state Markov process describing bacterial state transitions, the holding or dwell time in each state can take values of $1/\lambda_r = 0.1$ s for the tumbling state and $1/\lambda_t = 0.9$ s for the running state. Due to the independence property of exponentially distributed arrival events, for each attached bacterium at each time step (dt), a next arrival time is sampled from this distribution, via the inverse transform sampling technique, and decided whether the given bacterium will undergo a transition to running/tumbling in the next time step. This

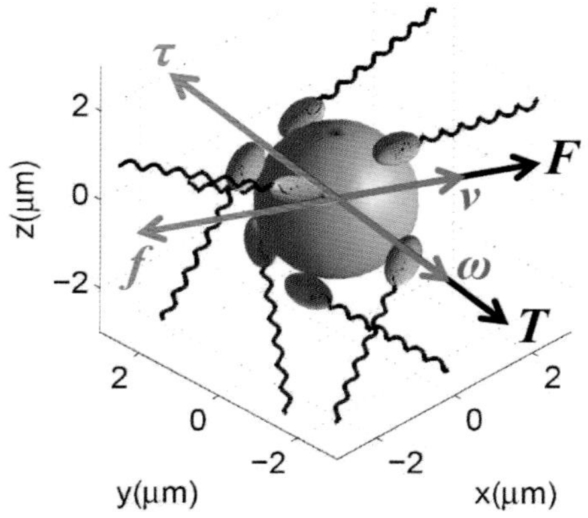

Figure 7.6

Free-body diagram of the swimming bio-hybrid microrobot, where $\vec{F}$ and $\vec{T}$ are the instantaneous total bacteria propulsion force and torque, $\vec{f}$ and $\vec{\tau}$ are the hydrodynamic translational and rotational drag on the bead, and $\vec{v}$ and $\vec{\omega}$ are the bead translational and rotational velocity vectors, respectively.

sampling is performed within a larger framework of an integration running forward in time.

Bacteria attached to the bead surface are modeled as point propulsive forces, having arbitrary initial orientation, position, and state. During the tumble state, the bacteria is assumed not to produce any propulsive force, while the run state yields a constant force of $F_b = 0.48$ pN from each bacterium [252]. Due to the Stokes flow regime, the beads inertial effects can be ignored, and hence its body dynamics are governed by the equations (see Figure 7.6):

$$\vec{f} = 6\pi\mu R\vec{v} = -\vec{F} = -\sum_{b=1}^{n} \vec{F}_b s_b \tag{7.8}$$

$$\vec{\tau} = 8\pi\mu R^3\vec{\omega} = -\vec{T} = -\sum_{b=1}^{n} \vec{r}_b \times \vec{F}_b s_b + 2(s_b - 0.5)\tau_b^m, \tag{7.9}$$

where b is the identifier of a bacterium, iterating from 1 to n (total number of attached bacteria), $\vec{F}_b$ and $\vec{r}_b$ are the propulsive force and position vector of each bacterium, s_b is a binary variable for each bacterium, being 1 during the run phase and 0 for the tumble phase, τ_b^m is the reactive motor torque of each bacterium on the bead, $\vec{f}$ is the hydrodynamic translational drag on the bead, $\vec{\tau}$

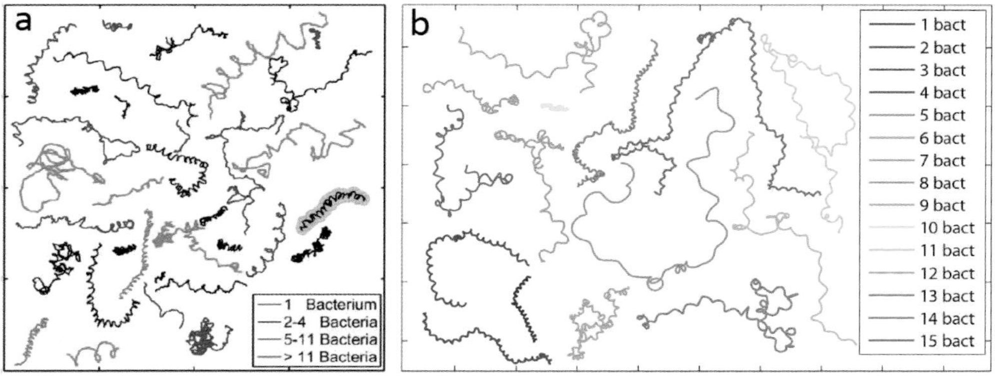

Figure 7.7

Sample 2D stochastic trajectories of bacteria-propelled beads with 5 μm diameter from (a) experiments (reprinted from [53] with the permission of AIP Publishing), and (b) simulations based on the computational stochastic bead motion model in Eqs. [7.8-7.9] for single and multiple (up to 15) attached *S. marcescens* bacteria.

is the rotational drag, $\vec{v}$ and $\vec{\omega}$ are the bead translational and rotational velocity vectors, respectively, μ is the dynamic viscosity of the aqueous medium, and R is the bead radius. During each integration step dt, the resultant linear and angular velocities of each bead, emerging from the contribution of all bacteria attached to it, are evaluated, and its motion is simulated forward in time. Random Brownian linear and angular displacements can also be included in the numerical model by allowing random displacement and rotation of the beads during each iteration characterized by root-mean-squared linear displacements of $\sqrt{2Ddt}$ in each of the three x, y, z directions, and angular displacements of $\sqrt{2D_R dt}$ around a random axis, where D and D_R are defined after Eq. [7.3] as the translational and rotational diffusion constants, respectively. However, the displacements caused by these processes are orders of magnitude less than those caused by bacteria propulsion and therefore are unnoticeable.

Using the previous computational stochastic bead motion model, simulations can be carried out for 5-μm-diameter beads propelled with a different number of bacteria with random alignment and position, and the simulation results in Figure 7.7(b) can be compared with the 2D experimental trajectories of the *S. marcescens*-propelled beads. It can be seen that the beads have the spiral type of trajectory behavior for especially few bacteria, but they show a great variety of different behavior when the number of bacteria increases.

Moreover, when the beads move near to the surface, this swimming behavior would change significantly [53] due to the fluidic wall effects in low Reynolds numbers (see Section 3.7.3).

7.3 Homework

Section 7.2.4 describes the design and modeling of bacteria-driven microswimmers in the Stokes flow regime (i.e., Re = 0). Let us have a spherical microparticle of radius R as the microswimmer body. Let us attach only one bacterium to the particle, where the elliptically shaped bacterium cell body attaches firmly to a random point on the surface of the particle with its long axis cell membrane in sideways to apply force and torque on the particle.

1. Answer the following questions for a bacterium always applying a constant propulsion force of 0.5 pN and a constant body torque of 4.5 pN·μm on the bead (see [53] and Figure 7.4c).
   a. Find the propulsion and rotational speeds of the particle as a function of the particle radius R.
   b. For $R = 2$ μm, find the propulsion and rotational speeds of the particle.
   c. For $R = 2$ μm, if there are two bacteria attached to the particle surface in parallel in the same direction in opposite sides of the particle, find the propulsion and rotational speed of the particle.
   d. For $R = 5$ μm, if there are 20 bacteria attached to the whole particle surface perpendicularly (heads up) with a homogeneous equal distribution and assuming the neighboring bacteria do not affect each other's fluidic propulsion, find the propulsion and rotational speeds of the particle.
   e. For $R = 5$ μm, if there are 10 bacteria attached to the half side of the (Janus) particle surface perpendicularly (heads up) with a homogeneous equal distribution and assuming the neighboring bacteria do not affect each other's fluidic propulsion, find the propulsion and rotational speeds of the particle.
2. Simulate and plot the random-walk 2D trajectory of a freely swimming bacterium with a run and tumble behavior as modeled in Eq. [7.7] with values of $1/\lambda_r = 0.1$ s for the tumbling state and $1/\lambda_t = 0.9$ s for the running state. Assume the average bacterium speed in the run state is 30 μm/s.

3. Simulate and plot the random 2D trajectory for a single bacterium propelling a spherical particle with $R = 1$ μm in question 1 with the stochastic bacterium run and tumble parameters in question (2) using Eqs. [7.8-7.9].

   a. Plot the mean-square-displacement (MSD) data of the bacterium-propelled particle for 100 sec.
   b. Find the rotational diffusion randomization time τ_R for the particle in water for 23°C temperature.
   c. For short durations of the MSD data with negligible rotational diffusion, fit the MSD data to find V_{mean} as defined in Eq. [7.5]. Show such MSD data and the fitted speed on the same plot.
   d. For long durations of the MSD data, fit the MSD data to find D_{eff} as defined in Eq. [7.6] and V_{mean}. Show such MSD data and the fitted speed on the same plot. Compare D_{eff} with the translational diffusion constant of the bare particle with no attached bacterium and discuss the difference with its reason. Is such a difference important for drug delivery applications if the particle was carrying drugs?

8 Remote Microrobot Actuation

This chapter covers the commonly used remote microrobot actuation methods. Remotely generated physical forces and torques can be used to actuate microrobots operating in a limited workspace, such as inside the human body or a microfluidic device. Main remote actuation methods based on magnetic, electrostatic, optical, and ultrasonic forces or pressures are explained below.

8.1 Magnetic Actuation

Magnetic actuation is widely used for remote microrobot power and control. Due to their ability to penetrate most materials (including biological materials), magnetic fields are naturally suited to control microscale objects in remote, inaccessible spaces. It is possible to independently apply magnetic forces and torques onto a magnetic microrobot using the magnetic field and its spatial gradient, leading to a wide range of microrobot design and actuation possibilities. As we will see, magnetic forces and torques can be relatively strong, offering the ability to do work not possible with other actuation schemes. In addition, several magnetic materials can be integrated with existing microfabrication methods. While magnetic effects have been observed by man for millennia, developments in magnetic materials continue today, driven primarily by the magnetic motor and digital recording industries. New magnetic materials have been discovered within the last several decades, allowing for more freedom in magnetic actuator design, and significant increases in the magnitude of magnetic forces that can be applied.

Magnetic forces and torques are applied to move a microrobot using fields created using magnetic coils or permanent magnets outside the workspace. The magnetic force $\vec{F}_m$ exerted on a microrobot with magnetic moment $\vec{m}$ in a magnetic field $\vec{B}$, assuming that no electric current is flowing in the workspace, is given by

$$\vec{F}_m = (\vec{m} \cdot \nabla)\vec{B} \tag{8.1}$$
$$= \begin{pmatrix} \frac{\partial \vec{B}}{\partial x} & \frac{d\vec{B}}{\partial y} & \frac{\partial \vec{B}}{\partial z} \end{pmatrix}^T \vec{m},$$

and the magnetic torque $\vec{T}_m$ is given by

$$\vec{T}_m = \vec{m} \times \vec{B}. \tag{8.2}$$

Thus, magnetic torques are generated from the magnitude and direction of the applied field, and act to bring a magnetic moment into alignment with the

Table 8.1

Typical magnetic material hysteresis characteristics. The first materials are referred to as magnetically "hard", while the last ones are "soft", and possess low remanence and coercivity

Material	Coercivity (kA/m)	Remanence (kA/m)	Saturation M_s (kA/m)
SmCo	3,100 [64]	~700 [64]	
NdFeB	620 [57]	~1,000 [64]	
Ferrite	320 [57]	110–400 [64]	
Alnico V	40 [57]	950–1,700 [64]	
Nickel	small	<1 [64]	522 [254]
Cobalt	small	<1 [64]	1,120-1,340 [255]
Permalloy (Ni-Fe)	small	<1 [64]	500-1,250 [64]
Iron	0.6 [57]	<1 [64]	1,732 [254]

applied field. Magnetic forces, however, are generated from the magnetic spatial field gradient, and operate on a magnetic moment in a less intuitive manner. As we shall see later, by controlling both the magnetic field and its gradients in the microrobot workspace, it is possible to provide independent magnetic torques and forces.

Magnetic materials can be classified as either magnetically hard, retaining their internal magnetization in the absence of a magnetic field, or magnetically soft, having an internal magnetization that is dependent on the applied field. Hard magnets, also called permanent magnets, are never truly permanently magnetized, but will behave as such until a large reverse field is applied to demagnetize them. Such a *coercive field* is typically much larger than the fields that are used to actuate microrobots, so the materials can often be treated as permanent. The strength of a permanent magnet after a saturating field is removed is referred to as the *remanence* and is an indication of the strength of the material. The coercivity and remanence of some commonly used magnetic materials are given in Table 8.1.

The magnetic strength of soft magnets depends on the magnitude of the applied field $\vec{H}$ as

$$M = \chi |\vec{H}|, \tag{8.3}$$

where χ is the magnetic susceptibility of the material. Susceptibility varies greatly with material, and is only constant up to the saturation of the material. At applied fields larger than the saturation field, the magnetization is constant.

For soft magnetic materials, the saturation magnetization can be a more relevant parameter than the remanence or coercivity, so these values are also given in Table 8.1.

Two mechanisms of magnetism are typically dominant in materials used for microrobots, namely, ferromagnetism and paramagnetism. The origin and study of ferromagnetism is complex, but it is the most common mechanism for most microscale magnetic robot materials. Some common ferromagnetic materials can be iron, cobalt, nickel, alnico, samarium cobalt (SmCo), and neodymium iron boron (NdFeB) and their alloys. In microrobotics applications, such materials are typically considered to be either perfectly soft or perfectly hard.

Paramagnetic materials are characterized as magnetically soft, and possess a low magnetic susceptibility χ. However, some materials, such as iron oxides behave as *superparamagnets* when in fine powder form, possessing large susceptibility. Such sub-micron particles can possess large magnetic moments, and are commonly used in small magnetic microrobots. For a detailed description of the origin and behavior of ferromagnets and paramagnets, the reader is referred to [64].

While magnetic forces do play a part in the motion of magnetic microrobots, at certain scales their effect is small when compared with magnetic torques. For example, for a permanent magnetic microrobot with dimensions $250 \times 130 \times 100$ μm^3 and magnetization of $M = 200$ kA/m (typical for a microrobot molded from rare-earth magnet material), the magnetic force a typical electromagnet can apply, with a gradient of $\nabla B = 55$ mT/m, is approximately $F_m = 36$ nN. By comparison, using Eq. [8.2], a magnetic torque of $T_m = 1.82 \times 10^{-9}$ Nm can be applied from the same coil with a field of $B = 2.8$ mT perpendicular to the direction of the microrobot magnetization. This torque, when treated as a pair of forces acting in opposite directions on the ends of the microrobot, acts as opposing forces, each approximately 7.3 μN. Thus, the effects of magnetic torques can dominate microrobot behavior at this size scale and are often used for actuation. In some such cases, the magnetic force can be neglected for analysis.

The units of magnetism are notoriously difficult to understand. Both SI and CGS systems are used in the literature, with different governing relations for each system. Units for some magnetic vector properties are given in Table 8.2. The magnetic field and magnetic flux density are often used interchangeably

Table 8.2
Units and conversions for magnetic properties [63] (to get SI units from CGS, multiply by the conversion factor)

	Symbol	CGS Units	Conversion	SI Units
Magnetic flux density	B	gauss	10^{-4}	tesla (T)
Magnetic field strength	H	oersted	$10^3/(4\pi)$	A/m
Volume magnetization	M	emu/cm^3	10^3	A/m
Magnetic moment	m	emu	10^{-3}	A m^2

in the microrobotics literature, and are related through the relationship

$$B = H + 4\pi M \quad \text{(CGS units)}, \tag{8.4}$$

and

$$B = \mu_0 (H + M) \quad \text{(SI units)}, \tag{8.5}$$

where $\mu_0 = 4\pi \cdot 10^{-7}$ H/m is the permeability of free space. Outside a magnetic material, $M = 0$, so the B and H fields proportional to each other. In some cases, B and H are thus used interchangeably for the magnetic field, although they do have distinct physical interpretations.

8.1.1 Magnetic field safety

For the use of magnetic fields penetrating the human body for remote actuation of mobile magnetic microrobots, the safety of high-strength magnetic fields can be a concern. However, the strength of a static magnetic field is not deemed dangerous up to 8 T [256, 257]. This large field threshold is not likely to be encountered in any microrobot actuation methods. However, time-varying magnetic fields can potentially pose a risk due to heating of tissue. The U.S. Food and Drug Administration (FDA) guidelines suggests (as a non-binding resolution) a range of safe amount of absorbed energy due to magnetic resonance imaging using large field changes over time, as measured by the specific absorption rate (SAR) [257]. For whole-body absorption averaged over a 15-minute period, this rate is 4 W/kg. Specific parts of the body, such as the torso or the extremities, can experience a larger SAR, as detailed in the report. The field rate of change to generate these SARs will depend on the tissue details as well as on the field strength and rate of change. FDA guidelines suggest that the field rate of change should be less than 20 T/s, which is three times less than the rate of change observed to stimulate peripheral nerves [258]. Many institutions, such as the Institute of Electrical and Electronics Engineers (IEEE), have published guidelines which generally limit the field rate of change to 0.1-1 T/s for frequencies up to 100 Hz [259]. These rates of change can be encountered in some of the magnetic actuation systems used in the microrobotics community. However, devices utilizing large fields are approved for medical use on a case-by-case basis using this report as a general guideline, as the exact conditions for operational safety depend on many other factors, such as body position and exposure time. Thus, devices using larger fields or field rate of change can be approved for use. Most of the regulatory reports have focused on MRI exposure, and may not be directly applicable for other medical devices. Thus, regulatory hurdles could be present if high rate of change magnetic fields are required for a medical microrobot application.

Of course, even small magnetic fields may be dangerous if the patient has pacemakers, surgical implants, etc. In addition, the use of very large permanent or electromagnetic coils can also have safety considerations in the operation room. Just like an MRI operation room, the use of magnetic materials in such environments should be restricted.

8.1.2 Magnetic field creation

Magnetic fields for actuating magnetic microrobots can be supplied by magnetic coils or large permanent magnets outside the microrobot workspace. Magnetic coils have the major advantage that they can deliver varying fields with no moving parts, and can be designed in a variety of ways to create spatially uniform magnetic fields and gradients. Permanent magnets, however, can provide large fields without the use of large electrical currents. The field in this case can be modulated by translating or rotating one or more external magnets, but in general cannot be turned off without moving the external magnets far from the workspace. We first discuss practicalities of generating magnetic fields and gradients using magnetic coils.

Magnetic coils are often designed to surround all or part of the microrobot workspace. The field created is typically assumed to be proportional to the current through the coils, an assumption valid if there are no nearby materials with non-linear magnetization hysteresis characteristics. The coil current I is governed by a differential equation (8.6), and depends on the voltage V_c across the coil, the coil resistance R_a, and inductance L_a as

$$\frac{dI}{dt} = \frac{-1}{R_a L_c} I + \frac{1}{L_a} V_c. \tag{8.6}$$

The control input is the voltage on the coil, and the current can be sensed using Hall effect current sensors if precise feedback control is required.

The magnetic field produced by a cylindrical coil is found by applying the Biot-Savart law for each turn over the path S as [260]

$$\vec{B}_{ec}(x,y,z) = \frac{\mu_0 N_t I}{4\pi} \int_S \frac{\vec{dl} \times \vec{a}_R}{|\vec{r}|^2}, \tag{8.7}$$

where $\vec{B}_{ec}(x,y,z)$ is the magnetic field at the microrobot's position (x,y,z) due to the electromagnets, N_t is the number of wire turns in the coil, $\vec{dl}$ is an infinitesimal line segment along the direction of integration, $\vec{a}_R$ is the unit vector from the line segment to the point in space of interest, and $|\vec{r}|$ is the distance from the line segment to the point of interest.

The principle of superposition holds for multiple field sources (assuming the workspace is free from soft magnetic materials), so the contributions from each coil can be summed to determine the total field. The full solution of Eq. [8.7] is given in [261], and here we give only the axial component of this solution for brevity. The flux density for N_t round loops of radius a, parallel to the $x-y$ plane, centered at $z=0$, is given by

$$\vec{B}_z = \frac{\mu_0 N_t I}{2\pi} \left[K(k) + \frac{a^2 - \rho^2 - z^2}{(a-\rho)^2 + z^2} E(k) \right]. \tag{8.8}$$

$K(k)$ and $E(k)$ are the elliptic integrals of the first and second kinds, respectively, which are given by

$$K(k) = \int_0^{\pi/2} \frac{d\theta}{\sqrt{1 - k^2 \sin^2\theta}} \quad \text{and} \tag{8.9}$$

$$E(k) = \int_0^{\pi/2} \sqrt{1 - k^2 \sin^2\theta d\theta}, \tag{8.10}$$

where

$$k^2 = \frac{4a\rho}{(a+\rho)^2 + z^2}. \tag{8.11}$$

A typical 2D magnetic coil setup is used in [40] to apply magnetic fields and gradients in-plane for 2D motion. This system uses orthogonal coil pairs to create these fields.

8.1.3 Special coil configurations

By pairing two coils along a single dimension, a special condition of spatial field or gradient uniformity can be obtained. Thus, the field can be assumed, within a small margin, to be invariant in space. To maximize the area of field uniformity, a Helmholtz configuration is used, where the space between the two parallel coils is equal to the coil radius [262]. By driving both coils equally in the same direction, a large region of uniform field is created between the two. This typically results in a nested configuration of the three orthogonal coil pairs, which is required due to the geometric constraints of using the Helmholtz configuration [263].

To maximize the area of field gradient uniformity, a Maxwell configuration is used, where the space between the two parallel coils is $\sqrt{2/3}$ times the coil radius and the coils are driven equally but in opposition to each other. It can be possible to combine independent Maxwell and Helmholtz coil pairs in one system to achieve both uniform fields and gradients.

8.1.4 Non-uniform field setups

Alternative coil configurations could have advantages when compared with the Helmholtz or Maxwell configurations, such as an increased level of controllable microrobot DOFs, at the expense of reduced areas of uniformity [16]. To calculate the fields and gradients created from a general coil system, we can use the following relations:

$$\vec{B} = \mathbf{B}\vec{I}, \tag{8.12}$$

$$\frac{\partial \vec{B}}{\partial x} = \mathbf{B}_x\vec{I}; \quad \frac{\partial \vec{B}}{\partial y} = \mathbf{B}_y\vec{I}; \quad \frac{\partial \vec{B}}{\partial z} = \mathbf{B}_z\vec{I}, \tag{8.13}$$

where each element of $\vec{I}$ is current through each of the c coils, $\mathbf{B}$ is a $3 \times c$ matrix mapping these coil currents to the magnetic field vector $\vec{B}$, and $\mathbf{B}_x$, $\mathbf{B}_y$, $\mathbf{B}_z$ are the $3 \times c$ matrices mapping the coil currents to the magnetic field spatial gradients in the x, y, and z directions, respectively. These mapping matrices are calculated for a given coil arrangement using Eq. [8.8] or by treating the coils as magnetic dipoles in space and are calibrated through workspace measurements as outlined in [1, 16].

Thus, using Eqs. [8.12-8.13] with Eq. [8.1] for a desired field and force on a single magnetic microrobot, we arrive at

$$\begin{bmatrix} \vec{B} \\ \vec{F} \end{bmatrix} = \begin{bmatrix} \mathbf{B} \\ \vec{m}^T \mathbf{B}_x \\ \vec{m}^T \mathbf{B}_y \\ \vec{m}^T \mathbf{B}_z \end{bmatrix} \vec{I} = \mathbf{A}\vec{I}, \tag{8.14}$$

where $\mathbf{A}$ is the $6 \times c$ matrix mapping the coil currents $\vec{I}$ to the field $\vec{B}$ and force $\vec{F}$. The equation can be solved if $\mathbf{A}$ is full rank, i.e., the number of coils c is greater than or equal to 6. The solution can be accomplished for $c \neq 6$ through the pseudo-inverse, which finds the solution, which minimizes the 2-norm of $\vec{I}$ as

$$\vec{I} = \mathbf{A}^+ \begin{bmatrix} \vec{B} \\ \vec{F} \end{bmatrix}. \tag{8.15}$$

If $c < 6$, then the solution will be a least-squares approximation. Having more than six coils leads to a better conditioned $\mathbf{A}$ matrix, which means a more isotropic workspace, reduction of singularity configurations, and lower coil current requirements. Systems designed to create such arbitrary 3D forces and torques have been created with six or eight coils arranged around the workspace in a packed configuration, first shown in [1] for moving magnetic seeds through the brain and more recently for microrobot actuation (the Octomag system) in [16]. The Octomag system is designed to provide easy access to one face, so all coils are on one side of the workspace. A similar but smaller system was shown in [264]. Another system with the coils completely surrounding the workspace is shown in Figure 8.1. The distance of each coil to the center of the workspace is the same for all coils, and they are arranged along the vertices of a cube, with the four lower coils rotated 45° about the z-axis to break symmetry (which would result in a singular $\mathbf{A}$ matrix). Here, two cameras view the microrobot from the top and side.

8.1.5 Driving electronics

Control of the currents driving electromagnetic coils is typically performed by a PC with a data acquisition system at a high control bandwidth up to several kHz, and the coils are powered by linear electronic amplifiers with optional Hall-effect current sensors for feedback. Coil currents of several A are typical, with higher currents used in conjunction with air or liquid cooling.

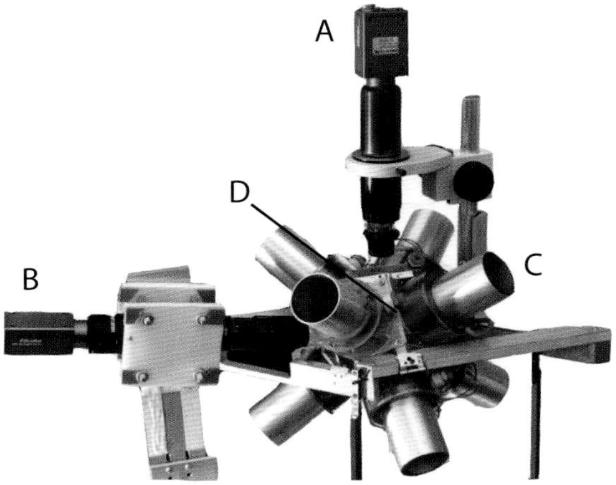

Figure 8.1

An example eight-coil system capable of applying 5-DOF magnetic force and torque in a several-cm-sized workspace with an uniform magnetic field. This system is capable of applying fields of strength 25 mT and field gradients up to 1 T/m using optional iron cores. A: Top camera. B: Side camera. C: Magnetic coils. D: Workspace.

8.1.6 Fields applied by permanent magnets

In some cases, it can be advantageous to use permanent magnets to create fields rather than electromagnets. Permanent magnets require no electrical power to generate and maintain a field, and thus are often well suited for creating large fields. However, modulating the field created by a permanent magnet requires moving or rotating the magnet. In addition, it is impossible in most cases to turn off the field created by permanent magnets, as is possible with electromagnets.

While complex changes to a magnetic field can be created by rotating arrays of permanent magnets in place, such as in a Halbach array [265], many systems for microrobotics use a single external permanent magnet, which is translated and rotated in space using a robotic actuator. Far from the external permanent magnet, the field can be approximated using the dipole model to simplify the calculations. The field $\vec{B}$ supplied by a magnetic dipole of moment $\vec{m}$ is given by

$$\vec{B}(\vec{m},\vec{r}) = \frac{\mu_0}{4\pi}\frac{1}{|\vec{r}|^5}\left[3\vec{r}(\vec{m}\cdot\vec{r}) - \vec{m}\,(\vec{r}\cdot\vec{r})\right], \tag{8.16}$$

where $\vec{r}$ is the vector from the dipole to the point of interest. While the dipole model is accurate in cases with a spherical external permanent magnet, the accuracy declines for other geometries and may contain a significant error when the distance from the magnet is small. As rectilinear and cylindrical magnets are widely available from commercial suppliers, optimal aspect ratios have been found for these designs [266]. For rectilinear shapes, the best aspect ratio is 1:1:1, or a cube, and for cylinders magnetized along the axial direction, the best diameter-to-length ratio is $\sqrt{4/3}$.

Many microrobot actuation schemes necessarily rely on rotating magnetic fields. A closed-form solution has been determined for the rotation axis of an external permanent magnet to create a desired magnetic field rotation at a point [267]. In this work, it was shown that a simple linear transformation can map the desired microrobot rotation axis to the required rotation axis of the external permanent magnet. Thus, the position of the external permanent magnet can be controlled independently of the required microrobot motion (with some constraints). Swimming microrobots have been shown to move using such methods, actuated by a single external permanent magnet tens of mm from the microrobot [268]. The permanent magnet in these cases is held by a multi-DOF robotic arm. Such results show that actuation using a single external permanent magnet could be sufficient for many clinical applications of microrobots inside the human body.

A related permanent magnet system called the Niobe (Stereotaxis) uses two very large permanent magnets moved with 3 DOF to create a field of approximately 80 mT over a space large enough to accommodate a human torso [269]. While this system is designed for steering magnetic catheters, it demonstrates that such permanent magnet systems are capable of being scaled up for clinical procedures.

8.1.7 Magnetic actuation by a magnetic resonance imaging (MRI) system

Magnetic actuation using a clinical MRI machine could leverage existing equipment infrastructure for navigating magnetic microrobots inside the human body. An MRI machine also has the potential to provide near-simultaneous microrobot localization in addition to propulsion.

Clinical MRI systems are designed for imaging and thus have several limitations for the propulsion of magnetic microrobots. Unlike the previously discussed magnetic coil systems, which can control the coil currents in each coil independently, an MRI machine provides a static field down the length of the system. This static field is provided by large superconducting magnets, and typically can be 1.5 T or higher, especially in MRI systems used for research. Thus the MRI system is well suited for the control of soft magnetic microrobots because this large field can saturate most soft magnetic materials. It could also be used with permanent magnetic materials, although the microrobot magnetization axis would be constrained to align with the static field direction. For imaging, magnetic field gradients up to ~40 mT/m can be created in any direction. These gradients can be used for microrobot propulsion by gradient pulling, and can potentially be increased through custom coil installations [47].

The excess heat created by the MRI system is also a major practical constraint. Because the systems are designed for periodic imaging only, they cannot provide large duty cycles at full field gradient generation. Thus, for continuous microrobot propulsion, the system must be operated below its maximum achievable gradient field capabilities [270].

It has been shown that the gradients required for navigation in difficult areas, such as the cardiovascular system, could be achievable using a clinical MRI system depending on the microrobot size [270]. In [271], it is suggested that custom gradient coils with a strength of 100-500 mT/m would be required to target tumors through the microvasculature using microparticles.

Due to these mentioned constraints, MRI-based control of microrobots in complex environments, such as the human cardiovascular system, requires sophisticated control algorithms. In [272], a high-level path planner is proposed to integrate the magnetic gradient steering and multiplexed feedback microrobot tracking. Such planning requires a detailed map of the environment, here acquired from pre-operative MRI images. A trajectory is planned using the Fast Marching Method, and a controller is developed to guide the microrobot in the presence of time-varying blood flow. Model-based control with adaptive algorithms [273] has the ability to increase the quality and robustness of microrobot tracking in such environments where instabilities and unmodelled dynamics consistently appear.

8.1.8 6-DOF magnetic actuation

Existing remotely actuated magnetic microrobots exhibit a maximum of only 5-DOF actuation, as typical microrobots have a uniform magnetization profile. In such robot designs, a driving torque about the microrobot magnetization axis

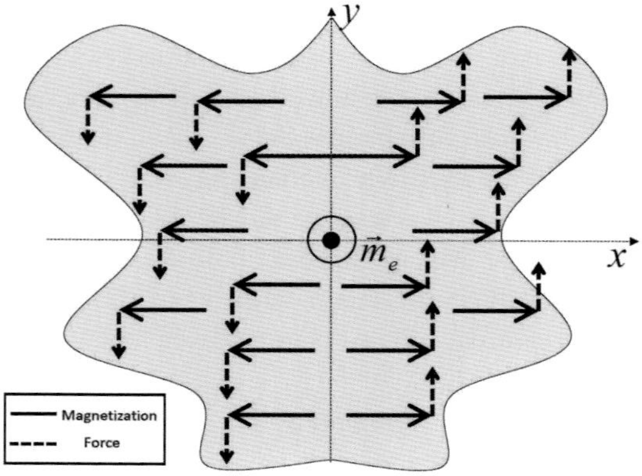

Figure 8.2

An example non-uniform magnetization profile to achieve 6-DOF magnetic actuation [31]. Copyright © 2015 by SAGE Publications, Ltd. Reprinted by permission of SAGE Publications, Ltd. Here, the robot has a net magnetization $\vec{m}_e$ along its local z-axis, and the magnetization vectors (solid vectors) are always pointing away from the origin. When a spatial gradient $\partial B_y/\partial x$ is applied, the induced forces on the magnetization vectors, indicated by the dotted vectors, exert torque around the z-axis of the robot body.

can not be achieved, limiting the DOF to five. This lack of full orientation control limits the effectiveness of existing microrobots for precision tasks of object manipulation and orientation for advanced medical, biological, and micromanufacturing applications. However, a non-uniform magnetization profile within the microrobot body can enable full 6-DOF actuation [31]. Such non-uniform magnetization allows for additional rigid-body torques to be induced from magnetic forces via a moment arm, as shown in Figure 8.2.

8.2 Electrostatic Actuation

A common alternative to actuation by magnetic fields is the use of electric fields. These can exert attractive or repulsive forces at small distances up to tens of microns for microrobot actuation and control on a 2D surface or at larger distances in an electrophoresis setup.

Localized high-strength electric fields are created by high-voltage electrodes under the surface. These electrodes can provide actuation power to microrobots through capacitive coupling, which is used to directly drive microrobots in

[38]. In [56], the substrate that microrobots move on has an array of independently controlled interdigitated electrodes to provide electrostatic anchoring for addressable multi-robot control. SU-8 is used as a barrier between the electrodes and the microrobot because it is inexpensive and has a high dielectric strength (112 V/μm), which will support the generation of the large electric fields necessary to anchor a microrobot without damaging the substrate. For experiments, a surface with four independent electrostatic pads was fabricated.

For the case of a conductive microrobot above an SU-8 insulation layer covering a set of interdigitated electrodes at an applied voltage difference of V_{id}, the conductor will assume a potential halfway between the two, or $V_{id}/2$, if it overlaps equal areas of electrodes at both voltages. With this assumption and considering negligible fringing, an estimate of the anchoring force F_{id} exerted by the interdigitated electrodes onto the microrobot is

$$F_{id} = \frac{1}{8} V_{id}^2 \frac{\epsilon_0 \epsilon_r}{t^2} A_{id}, \tag{8.17}$$

where A_{id} is the area of the electrodes overlapping the microrobot, t is the insulator thickness, ε_0 is the permittivity of free space, and ε is the relative static permittivity of the insulating material ($\varepsilon = 4.1$ for SU-8).

Electrokinetic forces can also be used to pull electrically charged microrobots in fields of several V/cm at cm-length scales. The actuation system contains a central workspace chamber surrounded by four large electrode chambers containing an ionic solution (Steinberg's solution) [39].

Millimeter and micrometer scale diodes can be propelled on water surfaces by converting an alternating current (AC) electric field into a local direct current (DC) field [274, 275]. Various types of miniature semiconductor diodes floating on water can act as self-propelling particles when powered by an external alternating electric field. The millimeter-sized diodes were shown to rectify the voltage induced between their electrodes [274]. The resulting particle-localized electroosmotic flow propelled them in the direction of either the cathode or anode depending on their surface charge. These self-propelled microrobots could emit or respond to light and could be controlled by internal logic. Diodes embedded in the walls of microfluidic channels can provide locally distributed pumping or mixing functions powered by a global external field. The combined application of AC and DC fields in such devices can allow decoupling of the velocity of the robos and the liquid and could be used for on-chip separations.

8.3 Optical Actuation

A focused laser light in different wavelengths can be used to remotely actuate microrobots by heating the focused region locally. Such local heating could create a mechanical deformation, force, and motion directly or change the surface tension of the liquid around the microrobot to propel it using the Marangoni effect. These two approaches are explained below.

8.3.1 Opto-thermomechanical microactuation

Microrobots with tens of μm size can be propelled on solid surfaces and steered using a focused laser light pulse that heats bimetallic mechanical structures of a microrobot locally to create mechanical impact force and motion. An example robot using such a technique consists of three-legged, thin-metal-film bimorphs designed to rest on three sharp tips with the robot body curved up off a surface [9]. Rapid, thermally induced mechanical bending (curvature change) of one of the legs due to the focused laser beam could lead to a stepwise 2D translation on a low friction surface. The laser parameters and focus location were used to control the velocity and direction of motion of the robot. Moreover, a microrobot body consisting of hydrogel bilayer sheets with a poly(ethylenglycol) diacrylate (PEGDA) layer bonded to a poly(isopropylacrylamide) (NIPAAM)-based graphene oxide nanocomposite layer can be folded by the heat of the focused near-infrared (NIR) light to encapsulate drugs or magnetic microparticles [276].

Microrobots can be driven directly by the transfer of momentum from a directed laser spot. For example, micrometer-sized objects with wedge shape, produced by photopolymerization and covered with a reflective surface, can reflect vertical laser light to the side, in the direction determined by the position and orientation of the objects [19]. The momentum change during reflection provides the driving force for the microrobots.

8.3.2 Opto-thermocapillary microactuation

A laser focused on or near a microrobot body on the water-air interface can locally heat the fluid interface and create a surface tension gradient spatially. As mentioned in Section 7.1.6, if the surface tension has spatial dependence along the interface, Marangoni flow would be induced at the interface. When the cause of the spatial variance of the surface tension is a local temperature gradient, the resultant phenomenon is called the thermocapillary effect. The thermocapillary stress (force per area), σ_t, due to the presence of such a stimulus is given by

$$\sigma_t = \nabla \sigma, \tag{8.18}$$

where σ is the surface tension in N/m. In 1D,

$$d\sigma = \gamma_t dT, \tag{8.19}$$

where γ_t coefficient denotes the susceptibility of surface tension to variations in temperature in N/(m.K) and is subject to the species of fluids in contact. Using such a relation due to laser heating,

$$\sigma_t = \gamma_t \frac{dT}{dx}. \tag{8.20}$$

This σ_t relation can be used to calculate the thermocapillary force per length, which could replace $\sigma = \sigma(x)$ term in Eq. [7.2] to calculate the Marangoni force due to optical heating.

Using the thermocapillary effect, a pulsed laser can be used to actuate an optically generated microbubble robot with 10-100 μm diameter on a solid substrate in 2D under water laterally with speeds up to 320 μm/s [277]. A single microrobot or a pair of microbubbles working in cooperation can be used to assemble polystyrene beads, single yeast cells, and cell-laden agarose microgels into different 2D patterns.

8.4 Electrocapillary Actuation

Electrical charges in a liquid-solid interface in air or other gases can be changed using electrical fields to control the liquid wetting properties locally and actively. Such an effect is also called *electrowetting* (see Section 3.2.2.1). In typical electrowetting systems, surface energy of an electrolytic fluid is modified using an electrode embedded in the substrate, which can result in motion of the fluid interface [278]. Such effects have long been observed, and have been used extensively in microfluidics [279, 280]. Recently, this effect has been demonstrated in a form like a mobile microrobot. In [41], a fluidic microrobot is formed from a water droplet trapped between two electrode layers and moved by electrowetting. By applying high voltages to embedded patterned electrodes in the supporting substrate, the hydrophobicity of one edge of the bubble can be changed, which results in horizontal forces on the bubble.

This microrobot can thus be pulled in 2D depending on the configuration of the embedded electrodes. While high speeds of up to 250 μm/s have been achieved with this system, the liquid nature of the microrobot renders it difficult to use as a contact manipulator, and it is restricted to operation in an air environment. However, the system naturally lends itself to parallel operation with multiple bubble microrobots working in tandem.

8.5 Ultrasonic Actuation

Ultrasonic waves are widely used to create medical images inside the human body safely with high penetration depths. Even, focused ultrasonic waves can be used for microsurgery or kidney stone removal as established surgical techniques. One approach is using ultrasonic waves to levitate biological or synthetic micro/nanoentities in vertical layers or in lateral specific patterns at the nodes of acoustic waves. For example, multilayers of mammalian cells are formed in a fibrin 3D microenvironment by the effect of bulk acoustic radiation pressure [281]. Due to their larger density and lower compressibility than the surrounding fluid, cells are driven to the node planes of acoustic standing waves, where there is minimal pressure. The interlayer distance is proportional to the frequency according to the equation $f = c/\lambda$, where c is the sound velocity and λ the wavelength. Thus, the acoustic frequency can be tuned to match the number and spacing of the interlayers, with an interlayer distance that corresponds to half of the acoustic wavelength. Moreover, 3D microtissues from cell spheroids is demonstrated by employing the vibration of acoustic standing waves and its hydrodynamic effect at the bottom of a liquid-carrier chamber [282]. A large number of cell spheroids could be assembled in seconds into a closely packed structure in a scaffold-free fashion under nodal pattern of the standing waves in a fluidic environment.

Use of ultrasonic waves in direct microrobotic actuation has not been explored much yet. As one recent work, remote ultrasonic waves can be used to propel microrobots by creating and ejecting microbubbles [283]. Ultrasound waves can rapidly vaporize a biocompatible fuel droplet (i.e., perfluorocarbon [PFC] emulsions) bound within the interior of a microrobot body for high-velocity, bullet-like propulsion. Gas and liquid PFC particles are biocompatible for intravenous injection and subsequent destruction upon ultrasound pulsing [20, 21]. The decreased solubility and low diffusion coefficient of these droplets and bubbles lengthens their blood circulation before an ultrasonic wave destroys or cavitates them [22]. PFC microbubbles or emulsions are thus attractive for diverse biomedical applications, such as externally triggered site-specific drug and gene delivery capsules and phase change contrast agents. Here, PFCs can be used as an integrated fuel source for microrobot propulsion.

When a remote ultrasound pulse (e.g., 44 μs, 1.6 MPa) is applied, metallic microtubes that contain trapped PFC droplets are propelled axially by rapid expansion, vaporization, and ejection of PFC bubbles. These robots (tens of μm in length) can reach a speed up to 6.3 m/s, and are powerful enough to penetrate deep into lamb kidney tissues. The speed and power of these ultrasonic microtubular robots can be modulated by the pulse length and amplitude of the ultrasound. A number of issues remain to be addressed for such a propulsion method to be useful in future biomedical applications. Although this method provides high power and speed, it currently relies on a finite supply of fuel and hence can operate for only a short time. Moreover, the robot size is hard to scale down to a few microns or smaller scale.

8.6 Homework

1. A permanent magnetic microrobot has dimensions of $200 \times 150 \times 50\ \mu m^3$ with magnetization of $M = 200$ kA/m.

    a. Compute the magnetic force that an electromagnet can apply on the robot with a spatial magnetic gradient of $\nabla B = 100$ mT/m.

    b. From the same coil, a field of $B = 5$ mT perpendicular to the direction of the microrobot magnetization can be applied. Compute the magnetic torque applied from this field on the robot using Eq. [8.2]. If this torque is created as a pair of forces acting in opposite directions on the ends of the microrobot, compute the force due to such torque on the robot. Compare this force with the one above and discuss whether applying a magnetic field or gradient is better to actuate a microrobot at this length scale.

2. Compare and list the pros and cons of creating magnetic fields and gradients using a permanent magnet versus an electromagnetic coil. How do such fields and gradients change as a function of distance in each case, and how would it affect the potential medical applications of magnetic microrobots?

3. For a spherical permanent magnet with radius R and magnetization of $M = 200$ kA/m, magnetic gradients can be used to pull it in 3D inside fluids. How much gradient is needed to pull $R = 5$ μm magnetic particle in a capillary tube filled with water with a 50 μm/s speed? Is such a gradient reasonable and safe to create by conventional MRI systems? How would the wall effects change the required gradient for the particle if the tube diameter is just slightly larger than the diameter of the particle?
4. What are the issues of using an optical remote actuation technique for a medical microrobot operating inside the human body? Which light source could penetrate the human tissue the deepest and how far for a safe operation?

9 Microrobot Powering

All current mobile microrobots have no on-board powering capability, and therefore they are typically actuated remotely or self-propelled by the fuels in the operation environment with no on-board functions, such as sensing, processing, communication, and computing yet. Only in the specific case of some bio-hybrid microrobot designs, the chemical energy (i.e., ATP) inside the cells powers the biomotors and thus the locomotion of microswimmers. Such on-board functions are indispensable for future medical and other microrobots with more advanced capabilities. Therefore, this chapter covers the possible on-board powering methods for microrobots: we can integrate an on-board energy/power source, transfer power wirelessly, and scavenge power from the operation environment.

Whichever on-board powering method is used, any small-scale robot or device would always have a limited power source. Therefore, all microrobot system designs should minimize their power consumption as much as possible. If we look at the power consumption components (locomotion actuation, communication, sensing, and computing) of a typical small-scale mobile robot system, the largest power-consuming component is the body locomotion actuation, which could be minimized with the proper locomotion mode design and body and actuator parameter optimization. For example, a rolling microrobot in air could have the least power consumption as the locomotion mode due to minimal friction forces resisting against the motion. For moving in a fluid, body drag forces should be minimized with proper body shape designs. Actively flying miniature robots in the air would have the worst locomotion power consumption, if they need to be actively levitated rather than passively levitated, such as in diamagnetic levitation [35]. Moreover, a locomotion actuation method for a given locomotion mode would have high or low power consumption as listed in Table 6.4 in Chapter 6 and other actuation-related chapters. Therefore, low power-consuming microactuation methods should be selected. Finally, actuation mechanisms with possible losses, such as locomotion and heat losses to the environment and internal friction, hysteresis and damping, could increase the consumed power, which also need to be minimized.

Sensors and microtools can potentially operate with much less power than microrobot locomotion actuation, and therefore could be viable candidates to receive power from stored or scavenged power. We now overview some potential on-board power sources, which could operate untethered at the microscale.

9.1 Required Power for Locomotion

To understand how much on-board locomotion actuation power (the most power-consuming portion of a microrobot) is required for a microrobot, we can have some basic calculations. The mechanical power required to move microrobots is dependent on the size scale and operation environment. For constant motion, the mechanical power P is equal to the force required for motion F times the velocity v as

$$P = Fv. \tag{9.1}$$

The estimated mechanical power required for stated values of several microrobots in the literature can be compared as shown in Table 9.1, using estimated force and forward translational velocity values from the literature. The Mag-Mite resonant magnetic microrobot [32], which is about 300 μm in size, operates at forces of approximately 10 μN with 12.5 mm/s maximum speeds, corresponding to an approximate power requirement of 125 nW. The Mag-μBot magnetic microrobot [13], which is about 200 μm in size, operates at forces of approximately 1 μN with 22 mm/s maximum speeds, corresponding to an approximate power requirement of 22 nW. The OctoMag magnetic microrobot [16], which is about 2,000 μm in size, operates at forces of approximately 83 μN with 1.9 mm/s maximum speeds, corresponding to an approximate power requirement of 340 nW. Moreover, another surface crawling robot SDA [284] uses electrostatic actuation from a powered micropatterned surface to move a 100-μm robot at up to 1.9 mm/s speeds in air, which require around 19 nW to move on the surface. Here it is observed that microscale surface crawling robots require tens or hundreds of *nanowatt* power.

For swimming microrobots at low Reynolds numbers (Re $\ll$ 1), such power consumption is related to the power spent to overcome the viscous drag. Using the relation $F \approx 6\pi\mu Rv$ in Eq. [2.4], the required swimming power for an approximately spherical body of a microswimmer is

$$P \approx 6\pi\mu Rv^2. \tag{9.2}$$

For example, a bacterium-inspired microswimmer [46] with 3- to 5-μm body size and 27- to 30-μm-long helical tail with speeds up to 3 μm/s, the required force can be estimated as around 0.2 pN, which means required power of 0.6×10^{-9} nW. Moreover, a spherical microparticle with 1-μm diameter that can be propelled by self-electrophoresis in H_2O_2 can reach a nominal speed of

Table 9.1

Maximum power required to move several mobile microrobots from the literature with their given actuation method, size, maximum speed, and estimated required maximum force to move

Microrobot	Actuation	Size (μm)	Forces (μN)	Max Speed (mm/s)	Power (nW)
MagMite [32]	Resonant magnetic	300	10	12.5	125
Mag-μBot [13]	Rocking magnetic	200	1	22	22
Octomag [16]	Magnetic pulling	2000	83	1.9	340
SDA [284]	Electrostatic	100	10	1.9	19
Swimming sheet [25]	Magnetic undulation	5900	1	100	100
Bacterium [46]	Magnetic rotation	30	0.2×10^{-6}	0.003	0.6×10^{-9}
Microswimmer	Self-electrophoresis	1	0.1×10^{-6}	0.01	1×10^{-9}

10 μm/s, which requires around 0.1 pN force and 1×10^{-9} nW power. Finally, a swimming undulating elastic sheet using remote magnetic actuation [25] with a length of 5.9 mm can reach 100 mm/s speeds, which requires around 1 μN force and 100 nW power. This level of power is quite low compared with macroscale robotic systems, but it is still too high to supply via on-board power supplies at the micron scale.

9.2 On-Board Energy Storage

Possible energy storage methods with their typical energy densities are listed in Table 9.2. Almost any of these methods may be adapted for use with microrobots, which still needs to be developed. Nuclear and chemical fuels could give the highest stored energy per volume for small-scale systems.

9.2.1 Microbatteries

Electrochemical cells transfer ions between anode and cathode materials across a conductive electrolyte. Microscale thin-film batteries have been made, although as battery storage capacity scales with volume, such storage techniques do not miniaturize well. The smallest electrochemical battery storage available in research stages are hundreds of nanometers thick thin-film construction, with energy densities of around 50 μA.h.cm$^{-2}\mu$m^{-1} and current densities around 10 μA.cm^{-2}, creating voltages of around 1.5 V. These cells have been fabricated in sizes of several mm. However, the integration and use

Table 9.2
Possible on-board energy storage methods with their estimated nominal energy densities (ATP: adenosine triphosphate)

Method	Energy Storage Density
Nuclear fuels, Uranium 235	1.5×10^9 kJ/L
Combustion reactants, gasoline	35,000 kJ/L
Electrochemical cells, Li-aV_2O_5	2,100 kJ/L
Thin-film batteries	1,500 kJ/L
Heat capacity, water at ΔT=20 K	840 kJ/L
Supercapacitors, lithium-ion	54 kJ/L
Fuel cells, H_2-O_2, 1 atm	6.5 kJ/L
Elastic strain energy	0.001-1 kJ/L
Magnetic field, 1.5 T	0.9 kJ/L
Electric field, 3×10^8 V/m	0.4 kJ/L
Molecular energy, ATP	30 kJ/mol

of these sources smaller than 1 mm in size for untethered operation has yet to be done [285].

The first sub-mm-scale microfabricated batteries were developed in 1996 [286]. Their cathode was $LiCoO_2$, electrolyte was LIPON, and anode was lithium. They were fabricated by RF sputtering and thermal evaporation of micropatterned nanofilms. Their cathode needs to be annealed at 700°C for high crystallinity and capacity. They could provide 3.9 V per cell and 150 mW/cm^2 specific power, and they were rechargable with up to 7,000 cycles lifetime. They are not integrated to any mobile microrobot yet.

9.2.2 Microscale fuel cells

The principle of fuel cells was discovered by Christian F. Schnbein in 1838. A fuel cell is an electrochemical device similar to a battery, but differing from the latter in that it is designed for continuous replenishment of the reactants consumed, i.e., it produces electricity from an external fuel supply of H_2 and O_2 as opposed to the limited internal energy storage capacity of a battery. The electrodes within a battery react and change as a battery is charged or discharged, whereas a fuel cell's electrodes are catalytic and relatively stable.

The basic operation principle of a hydrogen fuel cell is as follows (Figure 9.1):

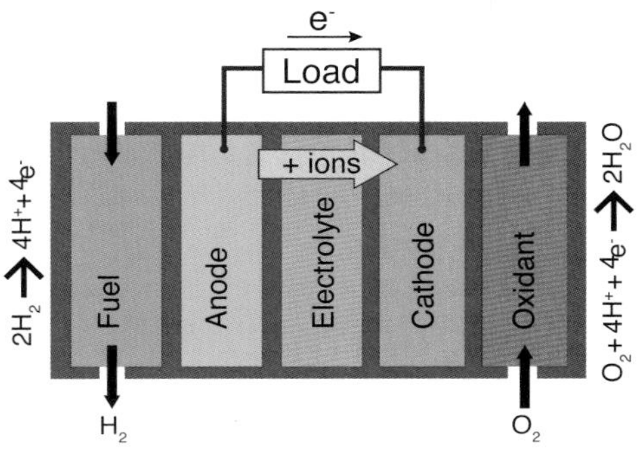

Figure 9.1
Basic operation schematic of a conventional fuel cell.

- Hydrogen diffuses to the anode catalyst, where it dissociates into protons and electrons.
- The protons are conducted through the membrane to the cathode, but the electrons are forced to travel in an external circuit (supplying power) because the membrane is electrically insulating.
- On the cathode catalyst, oxygen molecules react with the electrons (which have traveled through the external circuit) and protons to form water.

There are a wide range of fuel cell types, such as alkaline, polymer electrolyte membrane, direct methanol, biological, direct borohydride, formic acid, phosphoric-acid, solid-oxide, and zinc fuel cells. Polymer electrolyte membrane or proton exchange membrane fuel cells are the most common fuel cell types. They have low weight and volume, and they use a solid polymer as an electrolyte and porous carbon electrodes containing a platinum catalyst. They need only hydrogen and oxygen from the air and water to operate, and they do not require corrosive fluids like some fuel cells. They operate at relatively low temperatures (around 80°C), start quickly (less warm-up time), and are durable while a platinum catalyst is not cost effective.

In biological fuel cells, the electrode reactions are controlled by biocatalysts, i.e., the biological redox reactions are enzymatically driven, while in chemical fuel cells catalysts, such as platinum, determine the electrode kinetics. Microbial systems produce hydrogen as fuel for conventional fuel cells. Here, various bacteria and algae, e.g., *E. coli*, *C. butyricum*, *Clostridium acetobutylicum*, and *Clostridium perfringens*, are used to produce H_2 under anaerobic conditions [287].

Microscale fuel cells have been attempted by miniaturizing the device [288], while an H_2 fuel source still needs to be provided from a source, which is not typically that small. Microbial H_2 production might be promising for powering microrobots in the future.

As pros, fuel cells have high efficiency (40%-60%) and can be scaled down to micron scale if the fuel is provided from the environment or by microorganisms. However, as cons, providing H_2 fuel and water management are challenging, their voltage and efficiency drop with higher current, and they are hard to operate at low temperatures (must be above around 20°C).

9.2.3 Supercapacitors

Supercapacitors (or ultracapacitors) store energy electrostatically by polarizing an electrolytic solution. They consist of two non-reactive (carbon) nanoporous plates suspended within an electrolyte. Such plates have a high surface area to volume ratio to store a high volume of ions. Positive and negative plates attract the negative and positive ions, respectively. They have no chemical reactions and are highly reversible (charged and discharged hundreds or thousands of times). Typical commercial capacitance stored in supercapacitors is 4 F/cm^2 at 50 V. They can provide 4 kJ energy output at 15 seconds. Such an energy storage method is highly advantageous for small-scale robots that require high energy at short time scales, e.g., a jumping microrobot. After each discharging, they need to be recharged electrically. Although the principle is scalable down, current supercapacitors are centimeter scale. Sub-mm-scale versions need to be fabricated for specific microrobot applications.

9.2.4 Nuclear (radioactive) micropower sources

Another potential on-board power supply could be the harnessing of microstructured radioisotope power sources. Thin-film radioisotopes have been shown to exhibit extremely long half lifetimes of up to hundreds of years, with constant or pulsed power output and very large energy density. Such systems have been proposed for remote sensors [289], with typical energy density of 1 to 100 MJ/cm^3 but a constant power output of only hundreds of picowatts. The low power available could limit their application for microrobot actuation unless intermittent motion from charged storage could be utilized. In fact, direct mechanical motion through a bending MEMS cantilever has been used as a storage mechanism for intermittent signal transmission in such systems [290], which could be explored for microactuation purposes.

9.2.5 Elastic strain energy

For one-time powering, elastic strain energy can be stored inside a microrobot mechanism with elastic components, and such strain energy can move a mechanism when triggered on-board or off-board to move the robot, actuate an on-board microtool, or create electrical energy. As the major issue of this method, it is only for one-time use, and remote or on-board actuation is needed to repeat such mechanical energy storage.

9.3 Wireless (Remote) Power Delivery

To overcome the limitations with on-board electrical energy storage, electrical power can be remotely supplied by wireless delivery. This has been performed inductively, optically, or with microwave radiation.

9.3.1 Wireless power transfer by radio frequency (RF) fields and microwaves

Inductive power receivers can be made smaller than 1 mm using small pickup coils. These systems are driven by a transmitting coil with matched resonant frequency to the pickup coil. Using RF (2.4-2.485 GHz) power transfer, the power P_r received is equal to

$$P_r = \frac{P_0 \lambda_w^2}{4\pi R_{tr}^2}, \tag{9.3}$$

where P_0 is the transmitted power, λ_w is the signal wavelength, and R_{tr} is the distance between transmitter and receiver. Thus, in ideal conditions, the power delivered is proportional to $1/R_{tr}^2$ but in reality may decay even faster. Therefore, the distance between transmitter and receiver is a critical factor in such a design. For example, given a 1 W transmitter, the power received at the node 5 meters away would be 50 μW.

Such inductive power transfer has been used to deliver power up to tens of watts at distances up to several meters using large inductively coupled coils [291]. This work is significant in that it achieved high transmission efficiency of about 40%. However, these technologies are only beginning to be used in actual wireless power applications at short distances [292]. Indeed, as the power sent is proportional to the square of the receiving coil size, this technology will be difficult to scale down for use in sub-mm robots.

Using a pickup coil on the order of several mm, Takeuchi et al. [293] delivered several mW of power using a high-Q receiving circuit. The electricity delivered in this example powered an electrostatic actuator. However, such a design has not been minimized to the size required for sub-mm robots.

A related method is power transfer through microwave energy [294]. Using a rectifying antenna, power has been wirelessly transmitted since the 1960s with high enough efficiency to power free-flying helicopters [295]. This technology has more recently been under investigation to power ground vehicles and portable electronic devices. Its use in miniature actuators for robotic use has been investigated [296]. In a typical application, a centimeter-scale in-pipe robot was able to move at 10 mm/s using a supplied microwave power of 200 mW [297]. However, the scaling of microwave rectifying antennas could pose a problem for future miniaturization of the technology.

9.3.2 Optical power beaming

High-power laser beams can be focused on remote photovoltaic cells to gather light radiation and convert it to electrical potential. Such a dangerous power-beaming method is used in applications, where there is no human or other laser light-vulnerable biological organism or device involved in the operation environment. For example, a 9.2-mm diameter and 60-mm length cylindrical in-pipe inspection robot had 63 amorphous silicon photovoltaic cells connected in series at its back end inside a nuclear plant metal piping [294]. A laser light with 0.08 to 1 W at 532 nm was focused on the robot to induce 101 V and 88 μA electrical energy. Voltage-stabilizing circuits were used for the photovoltaic device to drive a MEMS electrostatic actuator on the robot.

9.4 Energy Harvesting

Energy can be scavenged from the microrobot's operation environment. Incident light, temperature gradients, pH, chemical fuels, mechanical vibrations and impacts, and air, wind, and fluid flows in the environment can be harvested to create electrical energy on the microrobot. Although the amount of harvested energy is small, it could be large enough for powering sensors and simple data transfer.

9.4.1 Solar cells harvesting incident light

Solar cells can be used to gather sun light radiation and convert it to electrical potential. The intensity of light from the sun outdoors is about 1.3 kW/m^2, although indoors the ambient light power density is only about 1 W/m^2. Most marketed solar cells are made from silicon (Si), although gallium arsenide (GaAs), amorphous silicon (a-Si), and cadmium telluride (CdTe) designs are also made. Efficiencies of Si, GaAs, a-Si, and CdTe solar cell materials are 24.5%, 27.8%, 12.0%, and 15.8%, respectively. A single solar cell creates a voltage potential of about 1 V, so arrays of such cells would be necessary to create the high voltages required for microrobotic piezoelectric or electrostatic microactuators. Solar cells have the advantage of being well-understood power sources, although limited power density (still high, 15 mW/cm^3 at clear sky outdoors [298]), low voltage output, and low availability of light in application areas could be major challenges. MEMS solar cells have been fabricated, with arrays generating tens of volts with μA-level current. An 8-mm walking robot was built, which used microscale 32 solar cells to power an electrostatic leg system for propulsion [299]. These cells provided approximately 100 μW of power in a $3.6\times1.8\times200$ μm^3 package with a mass of 2.3 mg. However, it could be difficult to miniaturize both the solar cell and associated electronics for sub-mm microrobotics use.

9.4.2 Fuel or ATP in the robot operation medium

The microrobots can harvest the liquid media that they operate as a chemical energy source if such media include any fuel that could be used to propel them. As we saw in the self-propelled microswimmers in Chapter 7, dilute hydrogen peroxide (H_2O_2), dilute hydrazine (N_2H_4), dilute I_2 or Br_2 solutions, glucose, and enzymes, such as urease and catalase, can be used as fuel to propel microswimmers with the proper surface materials. However, only several of such fuels (glucose and enzymes) exist in natural biological fluids. In addition to the fuels in the liquid media, ATP could also exist in the media, such as in biological fluids, which could be used as an energy source to actuate muscle and other cell-driven microrobots.

9.4.3 Microbatteries powered by an acidic medium

pH inside the liquid media that the microrobots operate can be harvested by integrating potato (lemon) microbatteries on the robot surface. For example, a potato battery is an electrochemical battery with Zn (zinc) and Cu (copper) electrodes inserted inside the potato with a distance. The acidic medium inside the potato acts as the battery buffer between the Zn and Cu ions. The Zn and Cu ions would still react if they touch within the potato, but they would only generate heat. Because the potato keeps them apart, the electron transfer has to take place over the electrical load element. Such a simple concept can be used in medical microrobot applications inside acidic media, such as stomach and urinary tract, and environmental microrobots inside acidic wastes, where Zn and Cu microelectrodes can be integrated on the microrobot surface to create electrical energy for the robot. As an example, 6 cm×3 cm-size urine-activated paper batteries gave around 1.5 mW average power output inside urine [300]. A Cu film, CuCl-doped filter paper and an Mg film were stacked and laminated between two transparent laminating films by passing through the heating roller at 120°C. Two slits are made to contact urine to the paper and remove the air from the battery.

9.4.4 Mechanical vibration harvesting

Energy scavenging can take advantage of natural mechanical vibration in the environment to generate electricity. This could be particularly well suited for high-vibration areas, such as on machinery or objects with which humans interact regularly. Many such vibration sources provide dominant frequencies between 60 and 200 Hz, with amplitudes from 0.1 to 10 m/s^2 [301]. Devices scavenge these vibrations using a free proof mass. This power can be harvested by piezoelectric [302, 303], magnetostrictive [304], electrostatic [305], or magnetic [306] elements. The electrical power available from a resonant system from oscillations of amplitude A_o is [307]

$$P_e = \frac{m\zeta_e A_o^2}{4\omega_o\left(\zeta_e + \zeta_m\right)^2}, \tag{9.4}$$

where ζ_e is the electrical damping ratio, ζ_m is the mechanical damping ratio of the oscillator, and ω_o is the oscillation frequency. Thus, it is seen that the power is proportional to the oscillator mass, the square of the oscillation amplitude, and inversely proportional to the frequency. Current centimeter-scale commercial piezoelectric vibration harvesters operate at 75 to 175 Hz and can provide 1 to 30 mW of electrical power. They can provide around 116 μW/cm^3 of power density [307]. Such harvesters need to be scaled down to sub-mm scale to provide power for microrobots on the order of few or less than μW.

9.4.5 Temperature gradient harvesting

Thermoelectric generators could be used to harvest electrical energy from thermal gradients [308]. The phenomenon of creating electric potential with a temperature difference (and vice versa) is termed thermoelectricity. On or inside the human body, outdoors, on computers, etc., significant temperature gradients can be harvested. Solid-state thermoelectric generators can provide around 40 $\mu W/cm^3$ of power density. However, these technologies may suffer from low energy conversion efficiency and difficulty in miniaturization.

9.4.6 Others

Air or liquid flow [309] in the environment can be harvested by microrobots for passive locomotion purposes. For example, aerosol-like neutrally buoyant microrobots could float in the air using air flow locally to monitor the environment. Also, inside blood, microrobots can navigate passively using the blood flow, and actively trigger a mechanism to move into a specific vein in any vein-branching region. Moreover, wind or fluid flow can be scavenged to create electrical energy on a microrobot with a proper device. Wind flow was used by a micro wind turbine in centimeter scale to harvest electrical energy [310], and energy was harvested from unsteady, turbulent fluid flows using piezoelectric generators [311].

Energy can be harvested from human motion. For example, during human walking, energy is dissipated during the shoe heel impact on the ground and shoe sole bending. By placing centimeter-scale polymeric PVDF and ceramic PZT flexible unimorph piezo beams inside front and back regions of the shoe sole, respectively, parasitic energy was harvested to power an RFID tag at some intervals. PVDF beam harvested on average 1.1 mW of power with 0.5% conversion efficiency. Parasitic power from pressure due to the heel strikes was on average 1.8 to 5 mW with 1.5% to 5% conversion efficiency because impact pressure was high and PZT material had better physical properties than PVDF for energy harvesting. However, PVDF was required in the shoe sole front region due to high bending stresses, where PZT would break easily in such stresses.

Earth's magnetic field, ambient radio waves [312], and many other environmental physical effects and chemicals could also be harvested as an energy source for microrobots.

9.5 Homework

1. Calculate the power requirement for the following microrobots at the Stoke's flow regime:

   a. A microrobot exerting force of around 10 μN to move in the given environment at 1 mm/s speed.

   b. A spherical microrobot with 20-μm radius swimming inside water at 50 μm/s speed. If the required speed of the swimming robot is doubled, how much power is required to propel it?

   c. A cubic microrobot with side length of 50 μm swimming inside water at 100 μm/s speed.

2. To minimize the required power to swim a microrobot inside a liquid, which shapes and sizes of the robot body would be the best?

3. List all of the possible on-board power sources for continuous operation of the following miniature robots:

   a. A milliscale flying robot in outdoors inspired by insects.

   b. An endoscopic capsule millirobot swallowed inside a stomach.

   c. A jumping or hopping milli/microrobot.

   d. A swimming milli/microrobot in acidic media.

4. To receive 10 mW of power on a millirobot that is 1 m away from an RF power transmitter (2.4 GHz), how much power needs to be transmitted? Is such power compatible with FDA regulations?

5. For harvesting electrical energy from surface mechanical vibrations for a spherical silicon microrobot with a radius of 200 μm crawling on a surface, what are the critical surface mechanical vibration parameters that would maximize the harvested energy, assuming we could control the surface vibrations directly? What is the typical range of transmitted power values for such a robot for typical environmental vibration sources with frequencies between 60 and 200 Hz and amplitudes from 1 to 10 m/s^2?

10 Microrobot Locomotion

Mobile microrobots, similar to biological cells and organisms, need to locomote (transport) from place to place through their environment using their actuators and possible motion mechanisms. They can have many different locomotion modes [313], such as:

- surface locomotion (crawling, rolling, sliding, walking, running, jumping, hopping, burrowing, and climbing),
- swimming (flagellar propulsion, pulling, chemical propulsion, body/tail undulation, jet propulsion, and floating),
- locomotion at the air-fluid interface (walking, jumping, climbing, sliding, sailing, floating, and running), and
- flying (levitated near-surface motion, floating, flapping wings, and rotary wings).

Future miniature robots would also have many of these locomotion modes simultaneously, such as jumping and gliding [314], swimming, crawling, jumping, and rolling, etc. similar to biological systems at the small scale to operate in unstructured environments with multiple and complex terrain (water layer on surfaces, obstactles in the environment, rigid ground, mucus-coated soft surfaces, etc.) [315].

It is crucial to understand the physics of each locomotion mode at the micron scale to optimize it for a given application and environment requirements. Main optimization objectives for microrobot locomotion are:

1. Maximize motion precision, speed, degree of freedom, and steerability.
2. Minimize energy consumption (cost of transport) by minimizing the robot speed and energy losses (drag, friction, damping, etc.).
3. Maximize stability so that the robot does not get stuck or fly away in an uncontrolled way.
4. Maximize endurance by minimizing wear, failure, sticktion, and fuel consumption.
5. Maximize robustness by minimizing sensitivity to environmental and robot parameter changes.

While optimizing these objectives, there are many constraints and compromises. Constraints could be limited actuation motion precision, speed, energy efficiency, lifetime, force/torque value and range, and any physical and geometrical constraints in the operation environment. As trade-offs (compromises), high robot speed and steerability would mean less motion precision, energy efficiency, endurance, and stability. For example, it is not possible to have a high speed and very low power-consuming microrobot both at the same time.

We will now study each locomotion mode with its given physical conditions, possible actuation methods, power consumption, and challenges. We will also give relevant biological counterparts for each locomotion mode.

10.1 Solid Surface Locomotion

A microrobot can move on the surface of a microfluidic device substrate, biological tissue, or ground in many different locomotion modes. For studying surface locomotion systems with periodic motion behavior, let us first define some important locomotion parameters. Stride is a complete cycle of movement, e.g., from setting down a contact point on the surface (e.g., foot) to the next setting down of the same contact point. Stride length (λ_s) is the distance traveled in one stride. Stride frequency is the number of strides taken in unit time. Mechanical cost of transport (T) is defined as the work required to move a unit mass of a robot (or biological organism) a unit distance such that

$$T = \frac{W_s}{m\lambda_s}, \tag{10.1}$$

where m is the body mass and W_s is the work done at a stride.

Froude number (Fr) is an important scale-independent (non-dimensional) parameter for surface locomotion, as discussed in Section 2.1, where it is defined as

$$Fr = \frac{v^2}{g\lambda_s}, \tag{10.2}$$

where v is the motion speed and g is the gravitational acceleration.

The most common surface locomotion mode at the micron scale is surface crawling in air or liquids. There are many different ways to crawl on a surface.

10.1.1 Pulling- or pushing-based surface locomotion

As the first and simplest surface locomotion, a microrobot can be pulled (or pushed) by an external propulsive force F_p laterally in air or liquids to slide with a constant speed v at the steady state. Physical condition for such sliding to start is

$$F_p > f_s, \tag{10.3}$$

where f_s is the static sliding friction between the robot and the surface, which is modeled in Eq. [3.72]. After the motion starts, if F_p is balanced with the kinetic sliding friction f_k (i.e., $F_p = f_k$), the robot would reach to a constant speed v at the steady state. If $F_p > f_k$, then the robot would accelerate with $a = (F_p - f_k)/m$, which is not desirable for precise and smooth motion control. At each stride, if the robot moves, reaches to a constant speed, and stops, the total work would consist of frictional (work done against surface friction), inertial (work needed to give kinetic energy for acceleration), and viscous drag (work done against robot body drag) works, such that

$$W_s = W_{friction} + W_{inertia} + W_{drag} = f_k\lambda_s + 2\frac{1}{2}mv^2 + bv\lambda_s, \tag{10.4}$$

where b is the viscous drag coefficient (e.g., $b = 6\pi\mu R$ for a spherical robot body) at a low Reynolds number, which depends on the properties of the fluid and the dimensions of the object. Then the total cost of transport is

$$T = T_{friction} + T_{inertial} + T_{drag} = \frac{f_k}{m} + \frac{v^2}{\lambda_s} + \frac{bv}{m}. \tag{10.5}$$

If the sliding friction is dominated by the load term and the adhesive term is relatively negligible, then $f_k \approx \mu_k mg$, where μ_k is the kinetic friction coefficient, and

$$T \approx \mu_k g + \frac{v^2}{\lambda_s} + \frac{bv}{m}. \tag{10.6}$$

Thus, the power consumption (cost of transport) for this locomotion method is dependent on μ_k, v, m, and λ_s. When the robot speed is high, $T \approx v^2/\lambda_s$. Thus, microrobots with higher speeds will consume quadratically more power to locomote, and smaller stride lengths (step sizes) would also increase the power consumption. In contrast, if the robot is slow, $T \approx \mu_k g + bv/m$, where μ_k and b will determine the consumed power. Therefore, b must be minimized by designing the robot body shape and size optimally.

Magnetic field gradients can be used to pull magnetic microrobots on surfaces using this method. However, as we see, such a method consumes high power at high speeds and could be jerky during fast motion starting and stopping, which makes them not too stable and precise. Therefore, it is not a common surface locomotion method.

10.1.2 Bio-inspired two-anchor crawling

Biological worm-inspired two-anchor crawling can also be used to move on surfaces. A robot body can be actuated laterally to lengthen and shorten periodically in a symmetric waveform while at the surface contact region of the body, there is a directional friction surface (e.g., angled bristles) to slide the robot easier and more efficiently in a forward direction rather than the reverse one. Thus, the robot would propel forward after one cycle of expanding and shortening.

Physical condition for two-anchor crawling is

$$\mu_b > \mu_f, \tag{10.7}$$

assuming the adhesion-based sliding friction term is negligible, where μ_f and μ_b are the forward and backward static friction coefficient of the directional friction surface, respectively. In this locomotion mode, the main work is done against friction at low speeds, assuming there are no other internal or external losses, such that

$$W_s = \mu_f mg\lambda_s. \tag{10.8}$$

Neglecting inertial work at low speeds, the frictional cost of transport is

$$T_{friction} \approx \mu_f g. \tag{10.9}$$

Thus, energy consumption for such a crawling system is determined by μ_f mainly.

At high speeds, inertial work done in the system can also become important. For simplification, we can think of the robot body as consisting of three equal parts [313]. For steady crawling at speed v, the middle part needs to have forward motion with speed v, and the front and rear parts need to be stationary at half the time and move forward with the speed $2v$ at the other half. Then

$$W_s = \frac{1}{2}\frac{2m}{3}(2v)^2 = \frac{4mv^2}{3}, \tag{10.10}$$

giving the inertial cost of transport as

$$T_{inertia} = \frac{4v^2}{3\lambda_s}. \tag{10.11}$$

When $T_{inertia} = T_{friction}$,

$$0.75\mu_f = \frac{v^2}{g\lambda_s} = Fr. \tag{10.12}$$

This means that the total cost of transport T is:

$$T \approx T_{friction} \text{ when } Fr \ll 0.75\mu_f, \tag{10.13}$$

$$T \approx T_{inertia} \text{ when } Fr \gg 0.75\mu_f. \tag{10.14}$$

Inertial cost of transport for crawling is proportional to v^2, which means higher speeds are not favored from the power consumption perspective. Maggots (larval Diptera) crawl similar to this model, where they have no bristles but use hooking to create directional friction.

10.1.3 Stick-slip-based surface crawling

Integrating directional friction surfaces on microrobot surfaces is not easy. Instead, low and high inertial forces can be created on non-directional friction surfaces by slow and fast contraction/extension cycles in forward and backward directions, respectively. During the slow contraction/extension, the robot sticks to the surface because the inertial force (impact) is not high enough to overcome the static friction on the interface. However, during the fast contraction/extension, the inertial force can be higher than the static friction, inducing sliding of the robot. Such stick-slip-based crawling can be implemented by microrobots easily, and thus it is a common microrobot surface locomotion method. However, modeling of such non-linear dynamic system is not simple, and we will study a rotational motion-based stick-slip locomotion method below to numerically compute the dynamics of an example stick-slip locomotion system.

10.1.4 Rolling

As we saw earlier, surface crawling methods consume significant power due to the frictional and inertial works done in the system. To minimize such works, rolling can be used as another surface locomotion method. For example, a spherical or cylindrical microrobot can be rotated around its center with a given torque to have a constant rotation speed, similar to a wheel. In a continuously rotating robot, the work is done against rotational friction at the micron scale (given by Eq. [3.78] in Section 3.6.2) and body drag in air/liquid, where the rotational friction is much lower than the sliding friction. Therefore, rolling is the most energy-efficient surface locomotion method, not only on the macroscale, but also on the micron scale. However, depending on a given application, the robot body shape and tool positioning requirements might not allow a rolling locomotion. Using magnetic field-based torques, spherical or cylindrical magnetic microrobots have been rolled on surfaces for efficient and fast locomotion [316, 317].

10.1.5 Microrobot surface locomotion examples

In the following, we define the implemented microrobotic surface locomotion approaches with their corresponding actuation methods and give a detailed case study on a rotational motion-based stick-slip locomotion method.

10.1.5.1 Magnetic actuation-based surface crawling

Magnetically actuated microrobots have been explored using a number of different approaches, spanning a wide length scale. These approaches use a combination of magnetic gradient pulling forces, induced torques, and internal deflections to achieve translation across a 2D surface. While magnetic forces can be used to move microrobots, such forces are relatively weak compared with those resulting from magnetic torques [81]. Thus, many actuation methods make use of strong magnetic torques for crawling in 2D. One major challenge addressed in these designs is overcoming high surface friction and adhesion. Many designs use a vibrating motion to periodically break the adhesion to allow for controlled motion with constant velocity. These approaches have accomplished fast and precise motion with full 3-DOF position and orientation control. Some of the methods are reviewed here.

The Mag-Mite system [32] uses low-strength, high-frequency fields to excite a resonant microrobot structure for smooth crawling motion. Speeds of tens of mm/s are achieved in an air environment, and lower speeds in liquid. The $300 \times 300 \times 70$ μm^3 microrobot consists of two magnetic masses, which are free to vibrate relative to each other, connected by a meandering microfabricated spring. The high-speed oscillation helps break static friction. Due to the small size of the microrobot, the resonant frequency of this oscillation is several kHz. The Mag-Mite is steered by applying a small DC field, which orients the entire microrobot in the plane. Motion reliability is increased through the use of a structured electrode surface, which adds asymmetric clamping forces to increase the directed motion amplitude. A version with increased biocompatibility is made using polymer springs with ferromagnetic masses using a simpler manufacturing process and similar performance [318].

A thin microrobot can be driven by the magnetostrictive response of certain materials. Magnetostriction is the internal realization of magnetic field-induced stress, analogous to the piezoelectric effect, which is electric field induced. Using terfenol-D as a high-strain material, several microns of steady-state deflection are created in a 580-μm microrobot. By driving with a 6-kHz pulsed magnetic field, a steady walking motion is achieved, with stated velocities up to 75 mm/s achieved in the NIST mobile microrobotics competition in 2010 [37]. Using controlled gradient fields, limited 2D path following is also achieved with this method.

A magnetic torque-based approach has been implemented, which allows for a simple magnetic microrobot to translate using stick-slip actuation, termed the Mag-μBot system [13, 319] (Figure 10.1). Due to the relatively strong magnetic torques, which can be created from small magnetic fields, this approach is robust to environmental disturbances, such as debris surface roughness or fluid flow at scales of tens to hundreds of microns. This method has been demonstrated in air, liquid, and vacuum environments, and thus is appealing for use in a variety of microrobotics applications. One major advantage is that the pulsed stepping motion results in small steps with a known step size. By regulating the pulsing frequency and angle sweep of each step, step sizes can be reduced to several microns. Additionally, the microrobot can be driven using large steps in conjunction with magnetic field gradients at speeds of several hundred mm/s, albeit with less precision control capability. This approach has also been used in [33] at a smaller length scale down to tens of microns. An even simpler method using magnetic torques is a rolling magnetic microrobot [34].

Case in study on rotational motion-based stick-slip surface crawling: As a case study, we will now study the dynamics of microrobot crawling motion using stick-slip dynamics, as presented in [13]. This locomotion method results in consistent and controllable motion in the presence of high surface adhesion forces, which tend to restrict microrobot motion. High-strength magnetic pulling forces can be used to move the microrobot, but in this case, the microrobot will experience high accelerations, leading to unpredictable behavior. To move reliably, a rocking motion is induced in the microrobot by using time-varying magnetic fields, which results in a controllable stick-slip motion across the surface using only magnetic torques, with no magnetic forces being required. In this method, the microrobot is rocked back and forth, with the microrobot angle being described by a sawtooth waveform. This results in a small slip to occur when the field angle is reduced quickly, and a sticking phase when the angle is increased slowly. This period is referred to as one step,

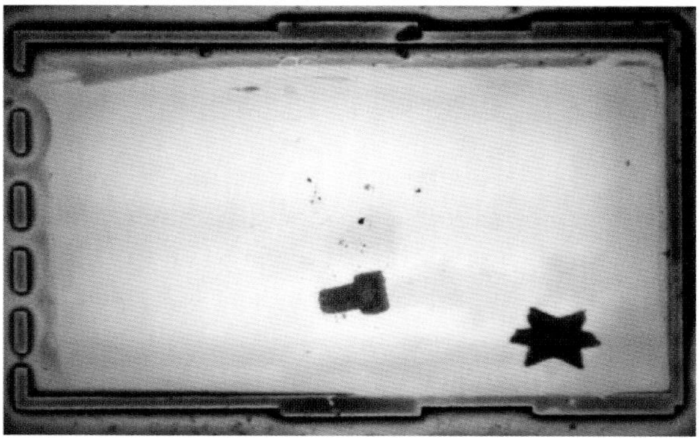

Figure 10.1

Top-view picture of an example 500 μm star-shaped Mag-μBot pushing a plastic peg into a gap in a 2D planar assembly task. The arena width is 4 mm.

and to study the motion in detail the microrobot dynamics must be studied in full. Because these dynamics are described by piece-wise functions when the microrobot is in contact versus out of contact with the surface, a computer numerical simulation is used to solve for the motion.

To simulate the dynamics of the magnetic microrobot in [13], only motion in the x-z plane is modeled, as shown in Figure 10.2. The robot has a center of mass (COM) at $\vec{X}$, an orientation angle θ measured clockwise from the ground, a distance r from its COM to a corner, and an angle $\phi = \tan^{-1}(H/L)$ determined from geometry. The robot experiences several forces, including its weight, mg, a normal force from the surface, N, an adhesive force to the surface, F_{adh}, an x-directed externally applied magnetic force, F_x, a z-directed externally applied magnetic force, F_z, a linear damping force in the x-direction, L_x, a linear damping force in the z-direction, L_z, an externally applied magnetic torque, T_y, a rotational damping torque, D_y, and a Coulomb sliding friction force F_f. F_f depends on N, the sliding friction coefficient μ, and the velocity of the contact point, $\frac{dP_x}{dt}$, where (P_x, P_z) is the bottom-most point on the microrobot (nominally in contact with the surface). Using these forces, we develop the dynamic relations:

$$m\ddot{x} = F_x - F_f - L_x \tag{10.15}$$

$$m\ddot{z} = F_z - mg + N - F_{adh} - L_z \tag{10.16}$$

$$J\ddot{\theta} = T_y + F_f \cdot r \cdot \sin(\theta + \phi) - (N - F_{adh})r \cdot \cos(\theta + \phi) - D_y, \tag{10.17}$$

where J is the polar moment of inertia of the robot, calculated as $J = m(H^2 + L^2)/12$.

In the simulation, the robot is first assumed pinned to the surface at (P_x, P_y), where $0 < \theta < \frac{\pi}{2}$. This gives the following additional equations:

$$x = P_x - r \cdot \cos(\theta + \phi)$$

$$\ddot{x} = \ddot{P}_x + r\ddot{\theta}\sin(\theta + \phi) - r\dot{\theta}^2\cos(\theta + \phi) \tag{10.18}$$

$$z = P_z + r \cdot sin(\theta + \phi)$$

$$\ddot{z} = \ddot{P}_z + r\ddot{\theta}\cos(\theta + \phi) - r\dot{\theta}^2\sin(\theta + \phi). \tag{10.19}$$

To solve Eqs. [10.15-10.19], we realize that there are seven unknown quantities $(N, \ddot{\theta}, \ddot{x}, \ddot{z}, \ddot{P}_x, \ddot{P}_z, F_f)$ and five equations, indicating an under-defined system. Because the stick-slip motion in this system is similar to the case outlined by Painlevé's paradox, we resolve the paradox by taking the friction force, F_f, as an unknown value (instead of assuming $F_f = \mu N$) [320]. Using the pinned assumption, we can set $\ddot{P}_x = \ddot{P}_z = 0$; then Eqs. [10.15-10.19] are solved analytically. Three possible types of solution can occur during each time step:

Case 1: The solution results in $N < 0$ (an impossible case). This implies that the pinned assumption was false, and the microrobot has broken contact with the surface. Eqs. [10.15-10.19] are resolved using $N = 0$ and $F_f = 0$, with $\ddot{P}_x$ and $\ddot{P}_z$ as unknowns.

Case 2: The solution results in $F_f > F_{fmax}$, where $F_{fmax} = \mu N$. This also implies that the pinned assumption was false, and the point of contact is slipping. Therefore, the robot is translating in addition to rocking. Eqs. [10.15-10.19] are resolved using $F_f = F_{fmax}$, $\ddot{P}_z = 0$, and $\ddot{P}_x$ as an unknown.

Case 3: All of the variables being solved for are within physically reasonable bounds. The robot is in contact with the surface at the pinned location and is rocking in place.

When a satisfactory solution for all seven variables is reached in each time step, these solutions for acceleration are used in the solver to determine the velocities and positions, which are in turn used as initial conditions in the next time step.

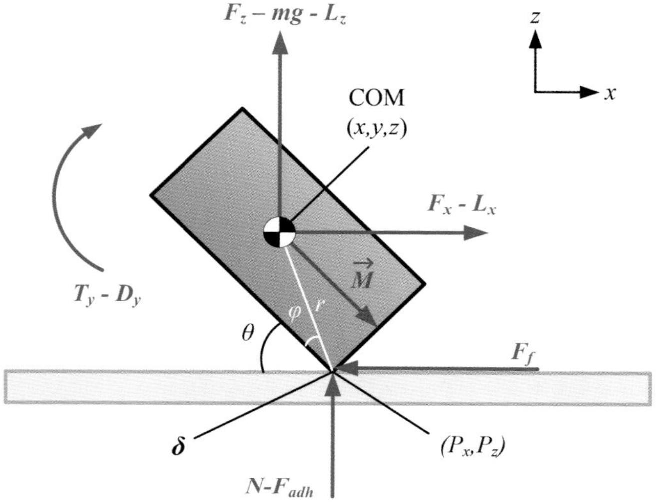

Figure 10.2

Schematic of a rectangular magnetic microrobot with applied external forces and torques. Here, the typical dimensions are several hundred microns on a side, and the microrobot is made from a mixture of NdFeB magnetic powder and a polyurethane binder. The magnetization vector is denoted by $\vec{M}$. The external forces include the magnetic force and torque F, T, the fluid damping force and torque L and D, the friction force f, the adhesion force F_{adh}, the weight mg, and the normal force N.

Numerical solver: To simulate the microrobot's motion, a fifth-order Runge-Kutta solver is used to solve the time-dependent system. A magnetic pulsing signal is given as a voltage waveform, and the magnetic field is solved for. With given initial conditions, the magnetic force and torque equations are used to determine the magnetic field forces, and Eqs. [10.15-10.19] are solved for the three position states of the microrobot: x, z, and θ. An example simulation case is compared with the experimental results in Figure 10.3. Here it is seen that the simulation and experimental results can match in some cases but vary widely in other cases, where the only change is a single material change. This highlights the sensitivity of the microrobot motion to microscale friction and adhesion parameters, which are difficult to measure accurately. Also note the relatively large error bars in the plot, which demonstrate the stochastic nature of the motion.

From this case study, we see that many different forces can play a significant role in microrobot motion, including friction and adhesion, fluid drag, and

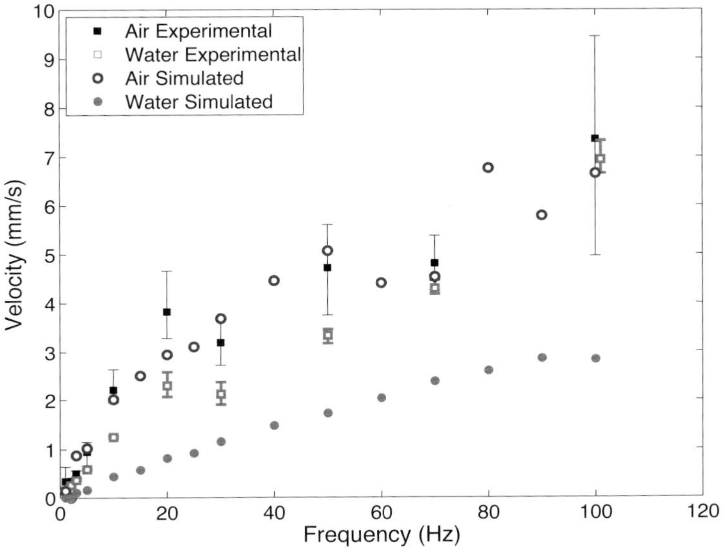

Figure 10.3

Example comparison of experimental and simulation microrobot speed values using stick-slip motion on a flat silicon surface [13]. Copyright © 2009 by SAGE Publications, Ltd. Reprinted by permission of SAGE Publications, Ltd. Average microrobot speeds are given for operation in two different operating environments: air and water. Error bars denote standard deviation in experimental results.

body inertia. The relative force magnitudes encountered in this case study are summarized in Table 10.1. The strong non-linearities associated with some of these forces highlight the need for analytical, numerical, or even finite-element solutions as indispensable design tools for studying and designing microrobot motions. To study the scaling of this particular motion method, we directly compare these physical forces for isometric scaling of the microrobot size in Figure 10.4. Here, we scale the microrobot size while keeping the actuating coils and coil-workspace distance constant because these will likely be fixed in a real microrobot application. We see that magnetic and fluid drag torques dominate the motion of the microrobot down to size scales of tens of microns. The relatively small size of the magnetic forces at sizes smaller than 1 mm motivate the use of microrobot propulsion methods, which utilize the magnetic torque.

Table 10.1

Approximate force magnitudes encountered in the magnetic microrobot stick-slip walking case study. Torques are treated as a pair of equivalent forces at opposite ends of the microrobot. For these comparisons, we assume a microrobot approximately 200 μm on a side with magnetization 50 kA/m, operating in a water environment on glass at a speed of tens of body-lengths per second in an applied field several mT in strength. Torques are treated as a pair of equivalent forces on opposite ends of the microrobot.

Force	Approximate Magnitude
Magnetic torques	1s of μN
Fluid damping torques	1s of μN
Friction forces	100s of nN
Normal forces	100s of nN
Adhesion	100s of nN
Weight	100s of nN
Magnetic forces	10s of nN
Fluid damping forces	1s of nN

Of interest when scaling below tens of microns in size is the increasing effects of thermal fluctuations, which lead to Brownian motion. As a rough indication of the strength of such forces, we can approximate the equivalent thermal forces by using the Stokes fluid drag equation (Eq. [3.86]), where we find the velocity using the average thermal energy relation $\frac{1}{2}m\bar{v}^2 = \frac{3}{2}k_B T$, where m is the mass, $\bar{v}$ is the average velocity, $k_B = 1.38 \times 10^{-23}$ J/K is Boltzmann's constant, and T is the temperature in kelvin. This relation thus will give us an indication as to the induced thermal fluctuation forces acting on a microsphere. When this force is calculated for a 1-μm diameter sphere of density 4,500 kg/m^3 at 293 K, we arrive at an equivalent approximate force of 1.6×10^{-11} N. However, a similar object slightly larger at 10 μm experiences an equivalent thermal force of only 5.2×10^{-12} N. Thus, it can be seen that such thermal fluctuations could dominate the motion of microrobots of several microns or smaller.

10.1.5.2 Surface crawling using light propulsion

Focused light energy can be used to remotely actuate a microrobot by heating or momentum transfer. These approaches require line-of-sight access to the microrobot, but they have some benefits, such as potential for simultaneous multi-robot control using multiple light sources.

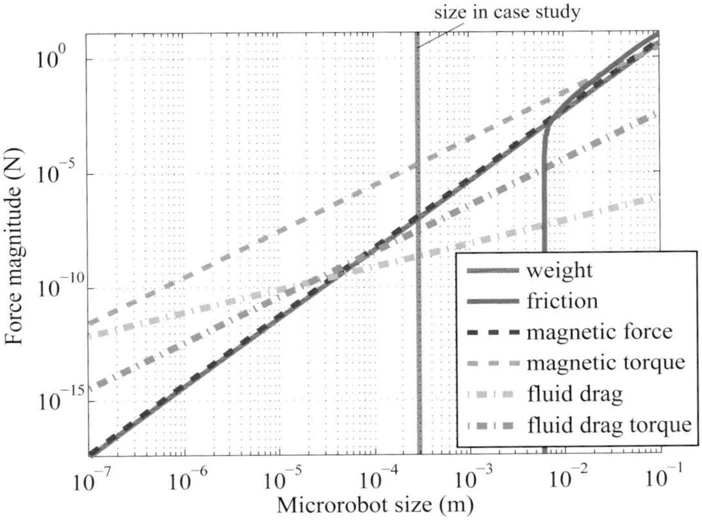

Figure 10.4

Mag-μBot force scaling from the case study. Equivalent forces are computed from torques by dividing by the microrobot size. The fluid environment is assumed as water with viscosity $\mu = 8.9 \times 10^{-4}$ Pa·s, and the microrobot density is 5,500 kg/m^3. The microrobot velocity is 1 mm/s, and its rotation rate is taken as swinging through an angle of 40° at a rate of 50 Hz, or about 70 rad/s. The magnetic field is taken as 6 mT and the field gradient as 112 mT/m, with a microrobot magnetization of 50 kA/m. To calculate surface friction, the interfacial shear strength is taken as one-third the shear strength, as $\tau = 20$ MPa, and the contact area is varying with load as given in Section 3.5. The gap size for adhesion calculation is taken as 0.2 nm. The coefficient of friction μ_f is taken as 0.41. The work of adhesion W_{132} is calculated in water for the polyurethane and silicon surfaces and is found to be negative, indicating repulsion. This material pairing was chosen specifically to yield this negative value. This results in a steep drop in the friction force when the microrobot weight overcomes this repulsive force at a microrobot size of about 7 mm. In a model with non-smooth surfaces, the friction would be positive at smaller scales.

In [9], a microrobot resting on thin metal bimorph legs is actuated by thermal expansion caused by a focused laser. By applying an asymmetric excitation, the dynamic heating and cooling behavior results in a forward step hundreds of nm in size. This can be applied with periods of several ms, resulting in directed crawling motion with speeds up to 150 μm/s. Because the motion relies on thermal gradients throughout the structure, a minimum device size of approximately 5 μm is stated. Below this size the heating covers the entire microrobot instead of being restricted to a single leg, resulting in no motion. Such a laser-based excitation method is limited to crawling on smooth 2D surfaces. In another approach, light pressure is used to push 5-μm wedge-shaped "sailboats" across a flat surface [19]. The driving pressure of approximately 0.6 Pa arises due to momentum transfer from the reflected light, and drives the microrobots at a speed of about 10 μm/s.

10.1.5.3 Electric field-based surface crawling

Electric fields are used secondarily in several microrobot actuation schemes, whereas some designs use the electrostatic force as direct actuation. In [284], a MEMS-designed scratch drive actuator is used to locomote on a 2D electrode surface in an air environment. Through careful design, an untethered actuator is achieved, which can be steered through the use of an integrated turning arm [8]. The actuator is 200 μm in width and consists of a flat plate suspended slightly above the substrate. When high-strength electric fields are applied, the plate bends downward towards the substrate. Through an asymmetric driving pattern, a stick-slip crawling motion is achieved, with step sizes less than 10 nm. By applying these steps quickly, speeds up to 1.5 mm/s are achieved. These scratch-drive microrobots are maneuverable in space when a high voltage is applied to the substrate, which pulls the turning arm into contact with the surface. In this reversible state, the microrobot rotates about the arm contact point.

A major limitation of electrostatic actuation is the necessary requirement for electrodes in the workspace. Because high field strengths are typically required, this could also limit applicability for biological or remote environments.

10.1.5.4 Piezoelectric propulsion-based surface crawling

Due to high levels of surface adhesion and friction in small-scale robotics, the breaking of this stiction is a major concern. One solution has been through the use of the high accelerations possible with piezoelectric materials. These materials experience strain in the presence of electric fields, and are typically driven by high-voltage potentials of several hundred volts. In [40], a lead zirconate titanate (PZT) piezoelectric element is integrated with a magnetic layer to form a hybrid microrobot.To actuate, a high-voltage impulse is applied between two electrode layers above and below the microrobot, and the generated strain in the PZT causes the microrobot to jump slightly, breaking the surface adhesion and momentarily reducing translational friction. High-strength magnetic field gradients supplied by magnetic coil pairs then act to pull the microrobot in the desired direction. Using this method, high translational velocities up to nearly 700 mm/s can be achieved, although precise control of microrobots at such high speeds is challenging. Because the dynamics of such actuation are fast, precisely modeling the behavior of this actuation style is difficult. Due to high accelerations and velocities, this method also may not be well suited to fine manipulation or assembly tasks.

More sophisticated use of piezoelectric elements in mechanical mechanisms, as is done at the milliscale [54, 321], requires on-board high-voltage power sources. Thus, miniaturization of such technology to the microscale could be challenging.

10.2 Swimming Locomotion in 3D

To move inside fluids in 3D, many swimming locomotion methods are possible: flagellar propulsion, pulling, chemical propulsion, body/tail undulation, jet propulsion, and floating/buoyancy. In all cases, to swim forward, robots and biological organisms must drive water backward. Swimming in low Reynolds number environments requires methods different from large-scale swimming. Because the first published in-depth study of the fluid mechanics of such swimming in 1951 [322], many fluid dynamics studies have been conducted to understand these propulsion styles, as reviewed in [323]. Primary means of swimming demonstrated at the microscale are inspired by biological methods: helical propulsion inspired by bacteria and flagella or body undulation inspired by spermatozoids. These methods are shown conceptually in Figure 10.5.

During swimming, the robot needs to be propelled in 3D inside liquids. Propulsive force F_p generated by a given swimming mechanism is balanced by the viscous (Stokes) drag forces on the robot body such that

$$T = T_{drag} = \frac{bv}{m} \tag{10.20}$$

for any swimming locomotion system at the micron scale. Also, the power consumption for such swimming propulsion would be

$$P = bv^2. \tag{10.21}$$

To minimize T and P, b must be minimized by designing the robot body shape and size optimally in a given Reynolds number regime, and v should not be high. Moreover, to minimize the power consumption, the robot body can be designed to be neutrally buoyant or the robot can be levitated passively

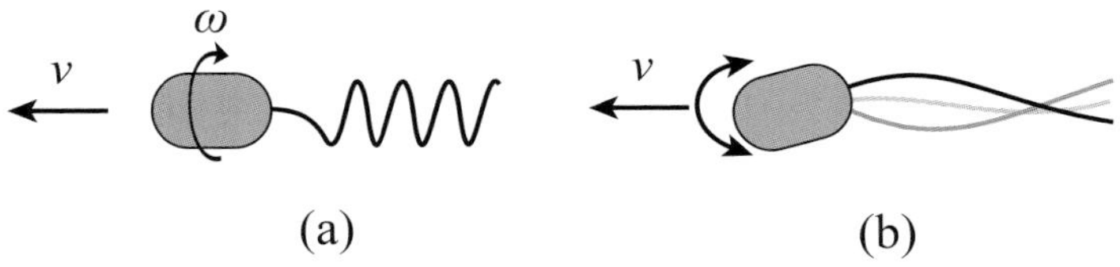

Figure 10.5

Microscale swimming methods. (a) Rotation of a stiff helix inspired by bundled flagella of swimming *E. coli* bacteria. (b) Traveling wave through an elastic tail or body inspired by spermatozoids.

using diamagnetic forces so that active propulsion is not needed to move the microswimmer in the vertical direction.

10.2.1 Pulling-based swimming

For motion in 3D, magnetic field gradients can be used to directly exert forces on a microrobot strong enough for levitation. Due to high accelerations possible with microscale components, high damping is often required to maintain control with this method. While larger robotic systems at the mm scale have achieved controlled levitation in air [324] using high-speed feedback control, microscale systems have been limited to operation in liquid environments. To manipulate a rigid-body magnetic microrobot in 3D levitation requires a high level of control. This is accomplished with a set of electromagnetic coils arranged around the workspace, which can simultaneously control the field and field gradient directions. Building on the work of Meeker et al. [1], a system capable of manipulating microscale robots uses eight independently controlled electromagnets, dubbed the OctoMag system [16]. Such a system can achieve 5-DOF control of a simple magnetic microrobot for levitation in liquid. The sixth DOF, rotation about the magnetization vector, is not controllable unless creative use of magnetic anisotropy in complex soft-magnetic shapes is considered [325]. In the OctoMag system, precise 3D position and 2-DOF orientation control of soft-magnetic or permanent microrobots is demonstrated in a workspace of several centimeters in a high viscosity silicone oil.

10.2.2 Flagellated or undulation-based bio-inspired swimming

In helical flagella-based swimming, a rotary motion activates a rigid helical-shaped body to propel through the viscous liquid. The fluid mechanics of helical swimming devices has been studied in depth, and the reader is referred to [326, 327] for a full review. In short, torque is generated in microrobotic helical swimmers using a magnetic head or tail and a uniform rotating magnetic field. The tail of such a swimmer is typically fabricated to be stiff and can be formed using stress-engineered curling thin films [11] and wound wire [328] and by glancing angle deposition [12] or microstereolithography [163].

The swimming force and torque on a rigid body are determined for a known rotation rate ω, applied driving torque τ, applied driving force f, and forward velocity v as [150, 329]

$$\begin{bmatrix} f \\ \tau \end{bmatrix} = \begin{bmatrix} a & b \\ b & c \end{bmatrix} \begin{bmatrix} v \\ \omega \end{bmatrix}, \tag{10.22}$$

where the matrix parameters a, b, and c can be determined for the shape geometry. For a helix with helix angle θ, these are given as

$$a = 2\pi n D \left(\frac{\xi_{\parallel} \cos^2\theta + \xi_{\perp} \sin^2\theta}{\sin\theta} \right), \tag{10.23}$$

$$b = 2\pi n D^2 \left(\xi_{\parallel} - \xi_{\perp} \right) \cos\theta, \tag{10.24}$$

$$c = 2\pi n D^3 \left(\frac{\xi_{\perp} \cos^2\theta + \xi_{\parallel} \sin^2\theta}{\sin\theta} \right), \tag{10.25}$$

$$\tag{10.26}$$

where n is the number of helix turns, D is the helix major diameter, r is the filament radius, and $\xi_{\parallel}$ and $\xi_{\perp}$ are the viscous drag coefficients found by resistive force theory parallel and perpendicular to the filament, respectively, as [330]

$$\xi_{\perp} = \frac{4\pi\eta}{\ln\left(\frac{0.18\pi D}{r\sin\theta}\right) + 0.5} \quad \text{and} \tag{10.27}$$

$$\xi_{|} = \frac{2\pi\eta}{\ln\left(\frac{0.18\pi D}{r\sin\theta}\right)}. \tag{10.28}$$

In magnetic rotational swimming, the typical microswimmers have a single rigid helical flagella. As another alternative, straight elastic, flexible multiple flagella have been used to propel microswimmers by magnetic body rotation [331]. Such flexible straight flagella would bend to have a curved shape during the body rotation, where the curvature amount can be controlled by the body rotation speed. Multiple flagella increase the total propulsion force and speed while they require more magnetic torque to actuate.

As a second method of swimming, traveling wave propulsion uses an elastic tail or body driven by an oscillating head affixed at one end, as seen in Figure 10.5(b). This oscillation is transmitted down the filament as a traveling wave, and the amplitude of the wave increases toward the end typically. Analyzing such motion requires solution of a coupled elastic-hydrodynamic interaction, and is thus quite complex. Approximate solutions have been found for small deformation with impulsive or oscillatory inputs [332] and by using numerical simulation [333], but no full analytical solution is available.

This type of actuation has been demonstrated at small scales using magnetic field actuation [334]. In [5], a crude traveling wave is induced in a 24-μm flexible filament attached to a red blood cell. Because the filament in this case is symmetric, the presence of the red blood cell acts to break the symmetry and allow for motion. It is also shown that other methods of breaking the symmetry of such swimmers is effective, such as the presence of a local filament defect [335].

In the case of an undulation of a deformable microrobot body-based propulsion, a milliscale swimmer used continuously rotating external magnetic fields to create traveling waves on the elastic and magnetic sheet body of the robot to propel it in liquids or at the fluid-water interfaces efficiently [25].

In [336], a detailed comparison of magnetically powered swimmers is performed, and it is found that helical and traveling-wave microsystems exhibit similar performance in terms of swimmer velocity for a given driving torque, and that both swimming methods compare favorably to magnetic gradient pulling as the microrobot size decreases.

10.2.3 Chemical propulsion-based swimming

Chemical reactions could be used in self-propulsion of microrobots in liquids, as described in Chapter 7. In [14], a microscale tube is used as a “jet” for motion in 3D through liquid. Propulsion comes from a stream of oxygen bubbles that form inside the tube through a catalytic reaction with the liquid medium. These 100-μm length tubes are made from layered Ti-Fe-Au-Pt, which rolls up passively due to residual stress. The Pt inner layer is allowed to react with the H_2O_2 solution to form oxygen gas inside the tube. Because the 5.5-μm diameter tube is naturally larger on one end, the bubbles exit from this end, and new solution is drawn in through the narrow end to feed the reaction. With frequent bubble ejection, speeds up to 2 mm/s are observed. By integrating a magnetic Fe/Co layer into the assembly, the orientation of the tube can be controlled using low-strength magnetic fields for steering in 3D.

The nature of chemical propulsion could lead it to be difficult to use for operation in arbitrary fluid environments. In addition, it could be difficult to harness such chemical reactions for sophisticated feedback motion control.

10.2.4 Electrochemical and electroosmotic propulsion-based swimming

Electric fields in a fluid can be used to create electo-osmotic propulsion for swimming microrobots. Such methods can be compatible with living organisms, and could be used in conjunction with other actuation methods. Electroosmosis puts to use a natural electrical diffuse layer, which surrounds any object in liquid. This layer is typically tens of nanometers thick, and contains a non-zero electric potential called the zeta potential ζ. In an electric field, the ions in this layer are pulled in the field direction. This motion drags the surrounding liquid, resulting in hydrodynamic pressure on the body, the speed of which depends on the zeta potential. A microrobot, which moves using this method, utilizes a large surface area to volume ratio to increase the propulsive force, but can be made in any shape. One example maximizes the surface area by using a helical shape, as shown in Figure 1.4(b) [43]. This swimmer is made of n-type GaAs, which adopts a negative charge in water. Using a 74-μm helix and a 240-V/mm electric field magnitude, a max speed of 1.8 mm/s was achieved.

Requirements for high-strength electric fields could limit the use of electrochemical or electroosmotic actuation in biological applications.

10.3 Water Surface Locomotion

Microrobots, like many tiny insects, can float, slide, and climb on water and other possible liquid surfaces. Statics and dynamics of possible water surface locomotion methods are explained below.

10.3.1 Statics: Staying on fluid-air interface

The statics of floating bodies residing at rest on the air-fluid interface is well understood. At the macroscale, buoyancy is the dominant force to lift the bodies on the fluid surface. However, such volumetric force with scaling relation of L^3 becomes almost negligible at the micron scale, and repulsive surface tension with scaling relation of L^1 can be used to generate lift for microrobots on the fluid surface. Bond number (Bo) determines whether buoyancy or surface tension (curvature force or capillary force) dominates the lift-generation mechanism [337], where

$$Bo = \frac{\text{buoyancy}}{\text{surface tension}} = \frac{\rho g h^2}{\sigma / w}, \tag{10.29}$$

where h is the mean depth below the unperturbed surface height, w is the width of the body in contact with the fluid, σ is the surface tension pressure, ρ is the fluid density, and g is the gravitational acceleration. At the micron scale, Bo $\ll$ 1 so that buoyancy is negligible.

Consider a body with density greater than that of the fluid $\rho_b > \rho$ and mass m floating at the interface (Figure 10.6). It would deform the fluid surface with a fluid meniscus with a contact angle of θ_c, and the lateral dimension of the deformed fluid is determined by the capillary length $l_c = \sqrt{\sigma/\rho g}$, which is around 2.6 mm for water. The body weight must be supported by some combination of the buoyancy force, F_b, and surface tension, F_{cap}, so that

$$mg = F_b + 2F_{cap} \sin\theta_c. \tag{10.30}$$

The buoyancy force is deduced by integrating the hydrostatic pressure $p = \rho g z$ over the body surface S in contact with the fluid, and so is equal to the weight of fluid V_b displaced above the body and inside the contact line, as shown in Figure 10.6. The surface tension may be deduced by integrating the curvature pressure over the same area. The surface tension is precisely equal to the weight of fluid displaced outside the contact line. The buoyancy and curvature forces are thus equal to the weights of the fluid displaced by the meniscus, respectively, inside and outside the contact line. Their relative magnitudes are the ratio of the characteristic body size w to the capillary length l_c. For thin microrobot bodies relatively much less than the capillary length (i.e., Bo $\ll$ 1), such as most water-walking insects, their weight is supported almost exclusively by surface tension. For macroscale large bodies, the vertical force is generated primarily by buoyancy.

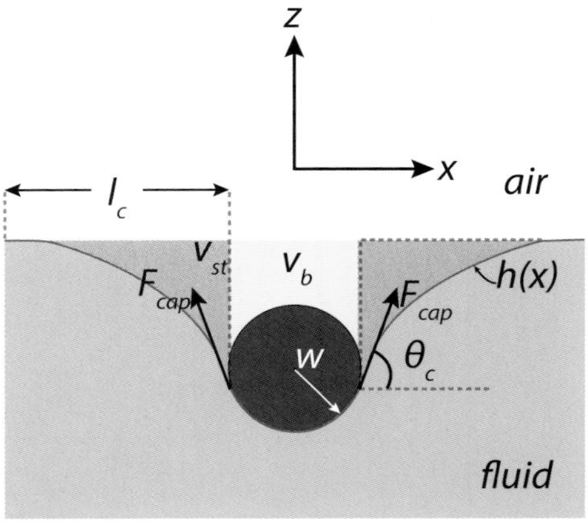

Figure 10.6

Side-view sketch of a cylindrical microrobot body with fluid contact angle θ_c staying on a fluid-air interface in equilibrium. V_b is the volume under water due to buoyancy, and V_{st} is the volume under water due to surface tension F_{cap}.

Assuming a cylindrical microrobot body, the Young-Laplace equation ($\Delta p = \gamma / r_k$) needs to be solved to compute the exact curvature of fluid $h(x)$ due to the surface tension [54, 338]:

$$\rho g h(x) = \frac{\gamma \ddot{h}(x)}{\left(1 + \dot{h}(x)^2\right)^{3/2}}, \tag{10.31}$$

where γ is the fluid surface tension (for water: 0.072 N/m), ρ is the fluid density, $\dot{h}(x) = dh(x)/dx$, and $\ddot{h}(x) = d^2h(x)/dx^2$. For given boundary conditions $h(x)$ can be solved, which can be used to calculate the volume V_{st} under the fluid level, e.g., lift due to surface tension.

Maximum lift force before the body breaks the fluid-air interface and sinks is dependent mainly on the body fluid contact angle θ_c of the body material and surface structure at the micron scale. Figure 10.7 shows numerically estimated lift forces by solving Eq. [10.31] for different θ_c values on a cylindrical body with length 20 mm and radius 165 μm. As we can see from this nonlinear behavior, it is better to have the body to be hydrophobic (i.e., $\theta_c > 90$ degrees) to maximize the body lift force. Having a superhydrophobic (i.e., $\theta_c >$ 150 degrees) body would improve the maximum lift force only slightly, so it is not crucial to have a superhydrophobic coating on the body. Water strider insects have wax-coated hairy legs, which are superhydrophobic. Although

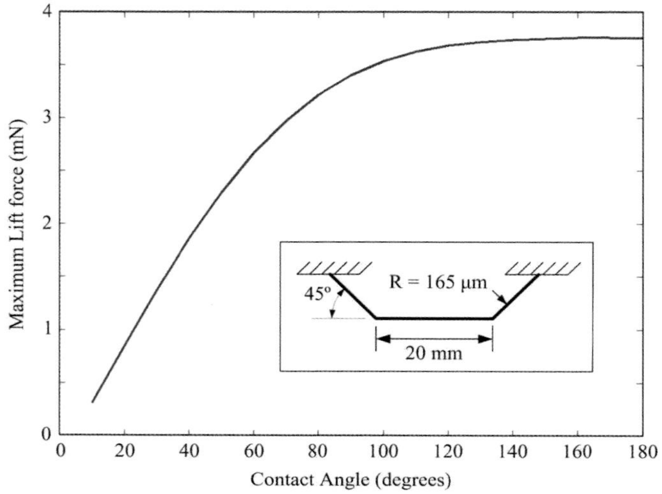

Figure 10.7

Numerically estimated maximum lift forces for different fluid contact angles (θ_c) of an example robot body with the cylindrical geometry shown in the inset image, showing that the robot body should be hydrophobic to have high surface tension-based lift forces [54]. Copyright © 2007 by IEEE. Reprinted by permission of IEEE.

such hairy legs do not improve the surface tension-based lift force significantly, they trap air and thus reduce the drag forces and help for jumping type of more impact-based locomotion on water with less water surface breaking issues due to higher impact resistance.

10.3.2 Dynamic locomotion on fluid-air interface

Possible dynamic propulsive forces on fluid-air interfaces could be viscous drag, form drag, added mass, surface tension, and Marangoni force [337, 339]. The form drag results from the pressure differential generated across the body at high Reynolds numbers. If the body strikes the surface asymmetrically, then it may utilize the hydrostatic pressure. The added mass force arises from the requirement that fluid be accelerated around an accelerating body, and the body's apparent mass increases accordingly. The surface tension can be used as a propulsive force, where fore-aft asymmetry in the meniscus of the robot body (i.e., when the total surface tension on the robot body has a finite lateral component) can generate their propulsion. Finally, Marangoni forces (i.e., surface tension gradients) through the release of surfactants or thermocapillary effects can propel microrobots on the fluid-air interface fast and efficiently. At low Reynolds numbers, the dominant propulsive forces would be viscous drag, surface tension, and Marangoni force while at high speeds form drag is also important.

Objects moving steadily at an air-water surface will generate surface waves if their speed exceeds the minimum wave speed, 22 cm/s. The wave field is typically characterized by capillary waves propagating upstream, and gravity waves downstream of the object. The waves radiate energy away from the robot body, and so represent a source of drag. Wave drag (the ratio of power lost through waves to the steady translation speed) is the dominant source of resistance for boats and large-scale robots moving on the air-water interface.

As an example of mainly surface tension-based water surface locomotion, some insects climb on water meniscus using surface tension. Wetting insects, such as *Collembola*, curl up their bodies so that surface tension pulls them up on the water meniscus [337]. Also, non-wetting insects, such as *Mesovelia*, climb the water meniscus using their specialized feet with retractable hydrophilic claws that allow them to grip the water free surface.

Water-strider insects move their two side legs with an elliptical trajectory to row on the water surface. They contact to the water surface without breaking it (normal leg force is less than the water surface breaking force of around 3.2 mN), move laterally, pull out from the water surface, and repeat such motion to propel using form drag and surface tension mainly. Many milliscale water strider robots have been proposed using such a locomotion principle. A tethered, 0.65-gr robot with six supporting stainless steel wire legs coated with Teflon, and two driving wire legs actuated by three unimorph piezo actuators at different resonant modes could propel the robot forward (with 3 cm/s speed) and steer it [54] (Figure 10.8(a)). Another untethered, 22-gr robot using two tiny DC motors, supporting legs with hydrophobic concentric circular feet, and on-board battery and electronics could create elliptical trajectories using its two driving legs to propel the robot with around 7 cm/s speed [111] (Figure 10.8(b)).

Required fluid-air interface locomotion forces are much smaller than a fully immersed swimming or surface crawling microrobot under fluids due to much reduced robot surface area and perimeter in contact with fluid; the most contact of the robot is with air, which has much less dynamic viscosity (around 50 times less than water) and density than fluids. Therefore, required actuation forces and power consumption are the lowest on fluid-air interface locomotion

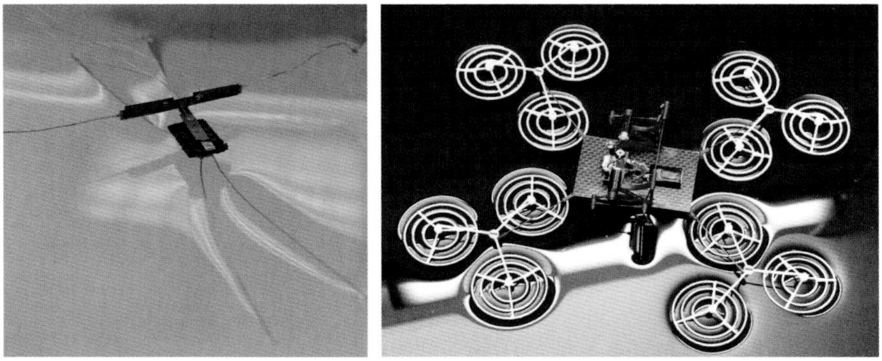

Figure 10.8

Example milliscale mobile robots that walk on water inspired by water strider insects: (a) a 0.65-gr tethered water walking robot with four water-repellent supporting legs and two driving legs actuated by three unimorph piezos at resonance (left photo), and (b) a 22-gr water walking robot with 12 water-repellent circular concentric feet and two driving feet actuated by two tiny DC motors (right photo).

systems, and many microrobots with low actuation forces locomote on fluid-air interfaces typically.

10.4 Flight

Active flight is the highest power-consuming locomotion mode because the robot needs to lift its body weight in addition to propel itself in air [340] during forward flight. Aerodynamics of flapping and rotating wings of flies, bees, moths, and hummingbirds at 30-1,000 Hz wing flapping frequency is unsteady and complex to model. A leading-edge vortex remains attached to the flapping wing as it translates during down- and upstrokes to generate lift, where such vortex leaves the wing at the tip creating a vortex ring behind. Approximate quasi-static aerodynamic models of translational, rotational, and added air mass lift have been commonly used to design such locomotion systems [165] while such models do not include wake capture or wing clap and fling type of possible unsteady effects. Scaled-up Robofly setup with the same Reynolds and Strouhal numbers with flies enabled many detailed insights and characterization parameters, such as lift and drag coefficients, for fly-inspired flapping wings [341].

To estimate the aerodynamic power requirement for hovering with flapping wings, the main physical condition is

$$mg = 2\rho l^2 \phi v_{ind}^2, \tag{10.32}$$

where m is the body mass, l is the wing length, ϕ is the wing flapping amplitude in radians, ρ is the air density, and v_{ind} is the induced air speed around the body generated by the flapping and rotating wings. Such a condition means that a tiny insect or robot needs to give enough momentum to the air to be able to balance its body weight. Then the induced power (the rate at which kinetic energy is given to the air) is

$$P_{ind} = 2\rho l^2 \phi v_{ind}^3, \tag{10.33}$$

which can be used to calculate the required aerodynamic power from the wing actuation system. For example, aerodynamic power density required for a few millimeter size *Drosophila* fly (Reynolds number: 100-250) to hover is estimated to be around 15 W/kg with around 11% energy efficiency. Robotic flies would have similar power density requirements to hover. To minimize the power consumption due to inertial work, many flying insects drive their wings close to the resonant frequency of the flapping wing system using elastic springs. At resonance, the wing system would do work mainly against aerodynamic forces.

Active flapping wings-based flying millirobots inspired by flies, bees, and hummingbirds have been proposed using piezo unimorph [209–211, 342–346], tiny DC motor [347, 348], electromagnetic [349], and piezo thin-film [350] actuators. These robots beat their wings at 10-200 Hz and mostly rotate their wings passively by using a torsional spring on the wing base joint. Due to the weight and size limitations, untethered insect-scale flying robots with on-board powering, electronics, and processing are not possible yet. Only untethered hummingbird or larger size scale flying robots have been possible so far.

Rotary wings-based flight inspired by mapple seeds and helicopters is another locomotion mode in air. One, two, or four rotary wings can be rotated at high speeds to lift and propel the body. Four rotary-winged 60-gr Mesicopter [351] and 75-gr microquadrotor [352], and one rotary-winged 200-gr robot [353] used tiny DC motors to lift off. However, such rotary wing systems are not favorable to scale down to few millimeter or smaller size scale while flapping wing systems are much more favorable.

Active or passive levitation can be used to lift a microrobot in the air at a given height from a surface. Active magnetic levitation can be used to suspend a few millimeter scale robots in the air [324], which requires significant power and control effort. In contrast, diamagnetic levitation can be used to passively levitate a microrobot [35]. As advantages of passively levitated microrobots moving along the surface laterally by magnetic forces, they have no surface friction and sticktion issues and can be fast, energy efficient, and precise.

10.5 Homework

1. Water striders use repulsive surface tension forces to lift their body weight statically and use side leg sculling motion to propel on water surface, as we discussed.

   a. What are the important design parameters for a water strider-inspired robot's legs and footpads?

   b. For a cylindrical wire leg design with 3-cm length, 0.2-mm diameter, and a surface coating with water contact angle of θ_c, calculate the lift force values for θ_c values of 30° and 100° using the simulated plot in Figure 10.7.

   c. Building water strider-inspired robots walking on water, as shown in Figure 10.8, why is a concentric circles-based footpad design used in the much heavier prototype? To lift a 80-kg adult statically on water surface, how many concentric circle footpads with 1-cm spacing among each footpad would you need? Is it meaningful to use such a lift principle for human scale?

   d. During propelling on the water surface, do water strider legs ever break the water surface (i.e., splash)? Which dimensionless number would determine whether such splashing would occur and what is the range of this number for water striders?

2. Another water surface locomotion system at the small scale is running on water with two legs, as in the case of basilisk lizards (see [339, 354–356]).

   a. Explain the physical mechanism of basilisk lizards running on water.

   b. What are the important design parameters and physical constraints for the lizard legs and feet and their motion to run on water without sinking?

c. Do these lizards break the water surface? Calculate the proper dimensionless number value range to decide such splashing for a given range of lizard weights and sizes.

d. How does the lizard stabilize its body motion while running on water [357]? Discuss the important physical and control parameters of the animal enabling stable water running.

e. Calculate the required leg rotation speed and power for a 50-kg athletic man to be able to run on water with a reasonably sized shoe size and shape.

f. If you want to design an untethered 100-gr robotic basilisk lizard, discuss the possible leg motion mechanism, foot material and shape, body design, actuators, sensors, and power sources that you would choose with the proper reasoning.

11 Microrobot Localization and Control

11.1 Microrobot Localization

Determining the location of untethered microrobots in a space is a major challenge, depending on the operational environment. Nearly all current microrobot control techniques rely on vision-based localization using conventional machine vision-automated tracking algorithms. Vision requires line of sight access to the microrobot workspace and may require more than one viewpoint to achieve 3D localization. For confined spaces, such as inside the human body, however, alternative localization techniques must be developed. As will be seen below, techniques to localize microrobots down to hundreds of microns in size pose significant challenges, although some concepts have been proven for tracking objects as small as tens of microns. Limits in microrobot localization capability could motivate the use of microrobot swarms, which could be easier to track in aggregate form.

11.1.1 Optical tracking

Optical tracking is possible for environments, which offer line-of-sight access to the workspace. Using one or more cameras fitted to microscope optics, the position of a microrobot can be obtained. Standard machine vision techniques, such as thresholding, background subtraction, edge detection, particle filters and color-space techniques [358], can be used to process an image in real time, providing position and potentially orientation information to the user or a feedback controller. More details were given in Section 11.2.

11.1.2 Magnetic tracking

Electromagnetic tracking: Electromagnetic tracking is possible using a paired magnetic field generator and sensor. Because the magnitude and direction of the generated field is position dependent, a field reading can be used to determine the position of the sensor relative to the field emitter. Such devices are available commercially in a tethered form (Aurora from ND, Flock from Ascension), with operating workspaces up to tens of centimeters in size. Because they depend on a precisely known magnetic field over the workspace, such devices are sensitive to the presence of magnetic materials in the proximity [359]. For increased sensitivity, these systems place the field sensor in the workspace, as the sensor can be made much smaller than the field emitter. An inverse setup could be possible, where the sensors are outside the workspace and the field generator is being tracked [360], but will suffer from small tracking range for a small field generator. Due to these challenges of low signal strength and magnetic distortions, there have been no examples of such wireless magnetic tracking at the scale relevant to untethered microscale robots. Enabling such a solution could require a significant advance in microscale remote field sensors on-board the microrobot or increased signal-to-noise ratio detection of the microrobot field.

Tracking by MRI: The clinical MRI machine is naturally suited to track the 3D position of microrobots [361]. If integrated with motion capabilities, the MRI machine could perform time-multiplexed localization and motion procedures for nearly simultaneous feedback control [47, 270, 272, 362]. The MRI image created also has the advantage of visualizing the structure of the entire workspace. For applications involving soft tissue, such as inside the human body, this could be critical information for navigation and diagnosis. Because MRI machines use magnetic fields for imaging, strong ferromagnetic microrobots can distort the local image, causing artifacts, which impede localization [362].

The MRI signature of microscale components containing magnetic and non-magnetic components has been studied. Using a ~150-μm cubic microcontainer, it was shown that the geometry and magnetic properties of the container can greatly change the resulting image [361]. However, through careful shielding, localization accuracies several times smaller than the object size could be obtained. Indeed, it has been shown that magnetic microrobotic elements much smaller than the imaging resolution of the MRI machine can be localized by analyzing susceptibility artifacts, tracking steel microspheres as small as 15 μm in diameter [363].

Thus, the MRI machine represents a useful tool for microrobot actuation and tracking studies as well as a potential infrastructure for future microrobot healthcare applications. However, the high cost of MRI machine operation could limit their appeal.

11.1.3 X-ray tracking

X-ray imaging has been used for medical imaging for many years and is particular adept at imaging objects with a unique density compared with their surroundings. X-ray imaging works by transmitting high-frequency electromagnetic waves through the workspace. An image is generated by sensing the attenuated signal after it passes through the workspace. In this way, it could be ideal for imaging microrobots moving in areas of soft tissue inside the human body.

Images in 3D can be generated from a series of X-rays taken in different planes using computed tomography (CT) scanning. Such 3D X-ray images typically have resolutions of 1 to 2 mm, while static X-rays have improved resolution of less than 1 mm [364]. Modern techniques could improve the resolution to several hundred microns [365]. Fluoroscopy uses an X-ray source to achieve continuous imaging with resolutions as high as several hundred microns using advanced detectors [366]. Thus, the use of X-rays for the localization of microrobots could be feasible and useful in certain applications.

A major downside in using X-ray imaging is the amount of ionizing radiation a patient is exposed to during imaging. This could limit its use in healthcare or other biological applications.

11.1.4 Ultrasound tracking

Ultrasound imaging is a low-risk alternative to X-rays for medical applications. It excels at localization in soft tissue and can provide frame rates over 100 fps or even higher for custom setups [367]. Ultrasound imaging works by transmitting a sound wave of several MHz and detecting the echoes to form an image. Ultrasound systems are commonly used, are low cost, and can easily provide accuracies better than 1 mm [368]. In general, a higher frequency operation yields better spatial resolution but less tissue penetration ability. One major challenge when using ultrasound is that it does not work well in the presence of bone or gas, and a skilled operator is required to operate and interpret the ultrasound images.

Passive ultrasound tracking has not been used in the localization of microrobots, but a milliscale device, which is remotely excited to emit ultrasound waves has been shown to result in high-resolution localization in a proof-of-concept demonstration [369]. This ultrasound emitter was excited remotely using high-frequency magnetic fields at approximately 4 kHz, and simulations suggest an imaging resolution of 0.5 mm could be achieved if the frequency is increased to 30 kHz, and the sensor placed 10 cm from the emitter.

11.2 Control, Vision, Planning, and Learning

Due to the inherently unstable nature of actuation by magnetic fields [370], feedback control of magnetic microrobotic systems is necessary to maintain a

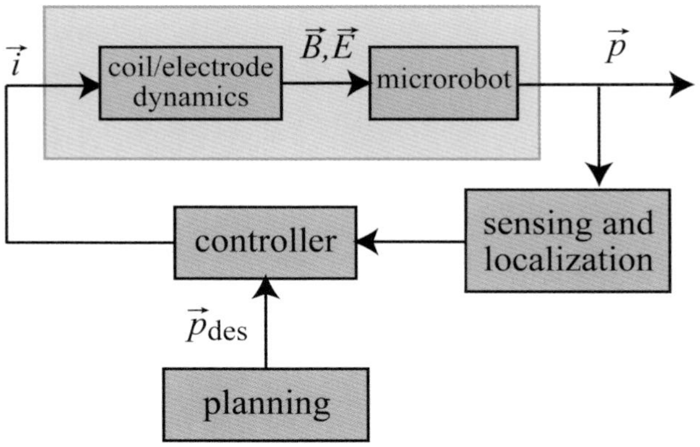

Figure 11.1
Components of feedback control of a general microrobotic system.

desired microrobot position or trajectory. A typical control system is shown in Figure 11.1. Here, the desired system input $\vec{p}_{des}$ is often a vector of position and orientation information, the size of which depends on the system DOF. The control system calculates a signal, which is sent to the coil or electrode system as $\vec{i}$. The coil/electrode dynamics result in an electric or a magnetic field $\vec{E}$ or $\vec{B}$ at the location of the microrobot. The position and possibly orientation of the microrobot is then observed using a microscope with machine vision or other localization scheme, which is fed back to the system controller.

Machine vision is often used to track the position of a microrobot. The task is to locate the microrobot in 2D or 3D space using one or more camera images in real time, and in the presence of optical noise and a cluttered environment. In some cases where background clutter can be controlled, relatively simple processing algorithms using thresholding and centroid finding are adequate to reliably locate a microrobot. However, in other cases, the addition of background subtraction, edge finding, dilation, particle filter algorithms [58], colorspace evaluation [358], and other methods are required [371]. Knowledge about the size, shape, and color of the microrobot can greatly aid in these processes, as in the feature-based tracking method (FTM). In FTM, features are determined from an object so that it can be recognized in any context. A scale invariant feature transform (SIFT) is then used to track the object under different image magnifications and rotations [372, 373]. A region-based tracking method is also commonly used when tracking with a relatively high frame rate,

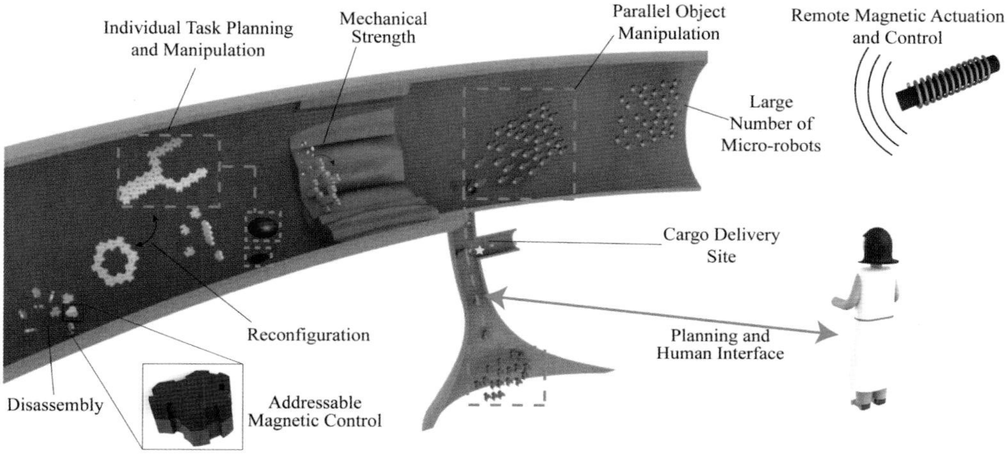

Figure 11.2

Conceptual sketch of a multi-robot control system, where a large number of magnetic microrobots could be remotely actuated and controlled or self-propelled autonomously to achieve a variety of tasks inside the human body or other operation environments [55]. Copyright © 2015 by IEEE. Reprinted by permission of IEEE. Here, such a microrobotic swarm could be addressed and controlled individually, as teams, or as an ensemble.

whereby tracked microrobots are assumed to be found in a location near the location found in the previous frame. As a further aid in microrobot localization, experiments can be performed in a low-clutter environment with high-contrast background and lighting.

One challenge when operating a microrobot in 3D space is that a single camera can only easily provide 2-DOF position information. This problem can be addressed by using two cameras arranged orthogonally to each other [16] or by using subtle information contained in the image from a single camera. Such clues can rely on the changing size and shape of a microrobot when it moves out of plane or relies on the predictable appearance of a defocused image, which depends on the out-of-plane distance [374, 375]. This single-camera 3D tracking has been performed in the presence of the optical distortion of looking through the lenses of the eye through compensation [376].

In practice, image processing and control feedback is typically performed at tens of Hz on a desktop computer system. Such low feedback rates are usually limited by camera frame rates and image processing speed.

11.3 Multi-Robot Control

One significant challenge in microrobotics is the control of multiple untethered agents as conceptually shown in Figure 11.2. Some microrobotic systems are naturally well suited for addressable multi-robot control, including those driven by focused light [9, 18, 277]. However, some of the commonly used actuation schemes, including control by magnetic or electric field operate remotely using a single global control signal. Thus, multi-robot control is difficult with these systems because driving signals are typically uniform in the workspace, so all agents receive identical control inputs. Without on-board circuitry and actuators to decode selective control signal, mechanical selection methods must be developed for the full control of multiple microrobots. Here we review some of the approaches used to address microrobots, which operate using a single global control input.

Researchers have shown the coupled control of multiple microrobots through the use of specialized addressing surfaces or through differing dynamic responses of heterogeneous microrobot designs, all in two dimensions on a flat operating surface. While some of these methods show promise for the distributed operation of many microrobots as a team, the limitation to operation on a 2D surface is significant as further developments in microrobots, especially for medical applications, will require 3D motion in liquid volumes.

Multi-robot operation on a 2D planar surface has been achieved in three ways: localized selective trapping, through the use of heterogeneous microrobot designs, and through selective magnetic disabling methods. Operation in 3D has been demonstrated through heterogeneous microrobot designs. Here we introduce these addressing methods and discuss their utility for potential distributed microrobot tasks using teams of independently controlled microrobots.

11.3.1 Addressing through localized trapping

In localized trapping, a spatially varying actuation is applied to only retard the motion of a single agent. This has taken the form of localized electrostatic [56] or magnetic [59] trapping and is capable of completely independent (non-coupled) control of multiple agents, at the cost of required embedded electrodes or magnets at a distance comparable to the spatial resolution of the addressing.

Motion of multiple magnetic microrobots has been achieved by employing a surface divided into a grid of cells, where each cell on the surface contains an addressable electrostatic trap capable of anchoring individual microrobots to the surface by capacitive coupling; this prevents them from being actuated by the external magnetic fields, as in Figure 11.3. This approach is related to that found in the field of *distributed manipulation* ([377]), where parts are manipulated in parallel using programmable force fields, but here the distributed cells provide only a retarding force while the actuation magnetic force is globally applied to all modules. For multiple microrobot control, the substrate upon, which the robot moves has an array of independently controlled interdigitated electrodes to provide selective electrostatic anchoring. For experiments, a surface with four independent electrostatic pads was fabricated.

The electrostatic clamping force provided by these electrodes was given by Eq. [8.17]. In short, the force is proportional to the square of the applied voltage and inverse square of the electrode-microrobot gap. In [56], such trapping forces are used with patterned electrodes to selectively trap magnetic microrobots, which move using the stick-slip crawling method from Section 10.1.5.1. Because the dynamics of this crawling motion involves complex surface interactions, it is difficult to predict the reduction in crawling speed in the presence of electrostatic trapping forces. In Figure 11.3, an experimental plot of microrobot velocity versus electrostatic anchoring voltage is shown, where it is seen that the required voltage is about 700 V to stop robot motion. Robot velocity does not monotonically decrease as voltage is increased, but experiences a local maximum near 550 V. For the purpose of multi-robot control, however, the critical voltage for effective anchoring is of importance.

This selective electrostatic trapping surface is useful as a potentially scalable method for multi-robot control in 2D. Some limitations include the requirement for high-strength electric fields, which may not be compatible with the manipulation of biological samples, and the limitation that all microrobots, which are not trapped move in parallel with the direction of applied magnetic field.

11.3.2 Addressing through heterogeneous robot designs

In addressing through heterogeneous robot designs, the agents are designed to respond differently to the same input signals. To achieve independent responses, where robot motions are not linearly related to one another, some

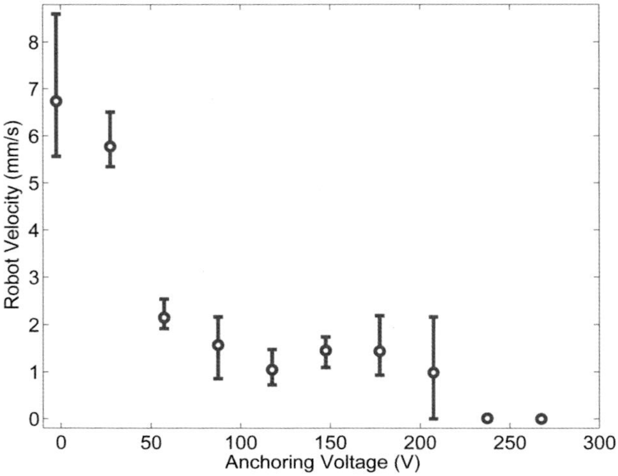

Figure 11.3

Robot velocity versus electrostatic anchoring voltage for a microrobot on a 6-μm-thick SU-8 layer. Reprinted from [56] with the permission of AIP Publishing. A critical voltage of 700 V is required to affix the microrobot. Videos of the motion were recorded and analyzed to determine velocities. A pulsing frequency of 20 Hz was used for translation.

type of dynamic response is required. In [38], different critical turning voltages are used to independently steer up to four electrostatically actuated microrobots, already introduced in Section 10.1.5.3. A similar approach uses microrobots with unique turning rates for differentiation [378]. Using appropriate control algorithms, independent positioning can be achieved with this method, albeit with limited control over the path taken. Turning arms are fabricated using stress-engineered MEMS techniques, to snap into contact at different critical voltages, which depend on the stiffness and geometry of the arm and its height above the electrode substrate. Unique voltages are created by changing these parameters, specifically the height and size of the arm. Microrobot actuation is accomplished through low-voltage stepping cycle, while arm state changes are accomplished periodically with a short applied V_{arm}, which ranges from 140 to 190 V. The snap-into-contact also exhibits hysteresis characteristics, which allows for more than two arms to be independently controlled by nesting the snap-down and snap-up voltages appropriately for each design. The actuation stepping voltage must then lie between the snap-down and snap-up voltages to allow for motion without altering the turning state of the microrobots.

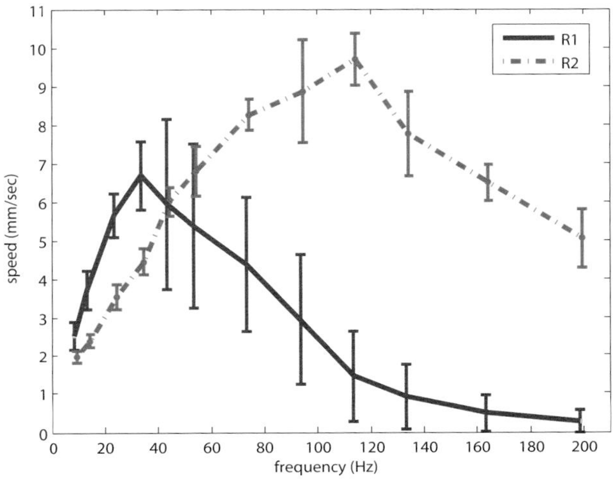

Figure 11.4

Experimental velocity responses of two Mag-μBots with varying aspect ratios but similar values of effective magnetization [45,67]. The maximum field strength was held at 1.1 mT. Data points are mean values, and error bars represent standard deviations for 10 trials.

In [379, 380], a dynamic stick-slip crawling motion is used to achieve independent but coupled velocity responses, as shown in Figure 11.4. This results in arbitrary positioning of up to three microrobots to goal positions in 2D, following a desired path within a span of a few body -lengths. Vartholomeos et al. [381] use a similar method, which relies on the non-linear drag of milliscale capsules with different sizes. This method could not be scaled down to the micron scale due to the reliance on inertial drag forces, which are negligible at smaller scales.

In [32], multiple resonant magnetic crawling microrobots are designed with different resonant frequencies, allowing for independent motion demonstrated on a specialized electrostatic surface. The frequency response of these resonant microrobots has relatively sharp peaks. Such independent addressing has been shown for two Mag-Mite microrobots using an associated electrostatic surface to aid microrobot motion.

When moving in 3D, available actuation techniques are reduced to swimming and direct pulling because there is no solid surface to push against as is used in the crawling 2D methods. To control multiple microrobot independently, the selective trapping method cannot be used because it relies on a nearby functionalized surface to provide trapping forces. Thus, the use of independent responses of heterogeneous microrobot designs is the only viable method for independent control in 3D. This has been accomplished for small groups of magnetic microrobots, which are pulled using magnetic field gradients [382]. Here, selection is accomplished by designing each microrobot to respond uniquely to rotating magnetic fields through different magnetic and fluid drag properties. This allowed for a unique magnetic force to be exerted on each microrobot, enabling an independent path following in 3D using vision feedback control. This remains the only experimentally demonstrated multi-robot control technique for 3D motion.

However, as another potentially viable method for microrobotic control in 3D, Zhang et al. [46] has shown the differing velocity response of unique artificial flagella with differing drag torque or magnetic properties but has not achieved independent positioning of multiple microrobots, presumably due to the inherent difficulties in controlling such swimming microrobots. Tottori et al. [383] have used an oscillating magnetic field to independently drive two artificial flagella swimming microrobots with different soft magnetic head designs. This also has not been used to achieve independent positioning of multiple microrobots.

11.3.3 Addressing through selective magnetic disabling

One method uses multiple magnetic materials with varying magnetic hysteresis characteristics in tandem to achieve addressable control. The magnetization of so-called "permanent" magnet materials in fact can be reversed by applying a large field against the magnetization direction, and the field required to perform this switch (i.e., the magnetic coercivity, H_c) is different for each magnetic material. For permanent magnetic materials, the coercivity field is much larger than the fields at which the microrobots are actuated for motion, allowing for motion actuation and magnetic switching to be performed independently. By using multiple materials with different magnetic coercivities, the magnetic reversal of each can also be performed independently by applying magnetic fields of the correct strength.

This independent magnetic switching can be used in microrobotic actuators to achieve addressable control of microrobotic elements. Our first addressable

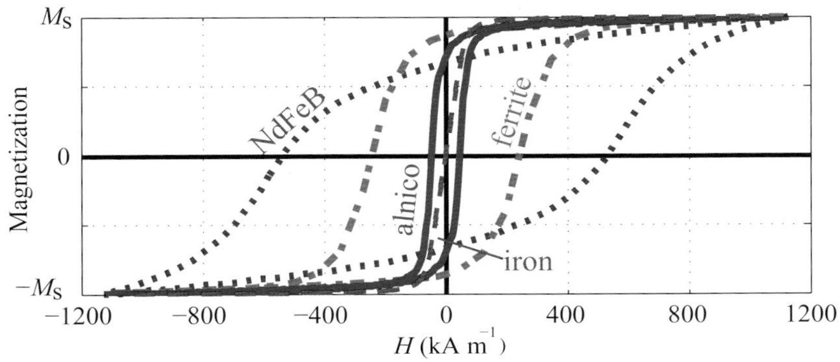

Figure 11.5

H-m hysteresis loops of microrobot magnetic materials, taken in an AGFM for applied field up to 1,110 kA/m shows distinct material coercivity values. The magnetization is normalized by the saturation magnetization M_s of each sample.

actuation scheme consists of several heterogeneous (each made from a different magnetic material) micromagnet modules interacting locally via magnetic forces. Selectively reversing the magnetization of one module can change the system from an attractive to a repulsive state. We present an experiment of this form, containing a set of heterogeneous magnetic modules floating on a liquid surface, which can be remotely reconfigured by application of a field of varying magnitude. In such a way, the morphology of the assembly can be altered arbitrarily into a number of states using a single applied field of varying strength. This implementation could be used for shape-changing microrobots, which adapt to the task at hand.

To achieve many-state magnetic control of a number of microrobotic actuators, we require a number of magnetic materials with different hysteresis characteristics [57, 384, 385]. The magnetic coercivity and remanence (retained magnetization value when the applied field H is reduced to zero) for a few commonly used materials are compared in Table 11.1, with coercivity values for ground powders measured in an alternating gradient force magnetometer (AGFM). In addition, the experimentally measured hysteresis loops for ground NdFeB, ferrite, alnico, and iron are shown in Figure 11.5. These materials cover a wide range of hysteresis values, from NdFeB and $SmCo_5$, which are permanent under all but the largest applied fields, to iron, which exhibits almost

Table 11.1

Magnetic material hysteresis characteristics [64] ([a]: measured in an alternating gradient force magnetometer (AGFM) after grinding)

Material	**Coercivity** (kA/m)	**Remanence** (kA/m)
SmCo	3,100	~700
NdFeB	620[a]	~1,000
ferrite	320[a]	110–400
alnico V	40[a]	950–1,700
iron	0.6[a]	<1

no hysteresis. For comparison, the magnetic fields applied to actuate magnetic microactuators are smaller than 12 kA/m, which is only strong enough to remagnetize the iron. Thus, the magnetic states of SmCo, NdFeB, ferrite, and alnico can be preserved when driving an actuator. This can be used to independently control the magnetization of each material, even when they share the same workspace. By applying a series of pulses in the desired direction greater than the coercivity field (H_c) of a particular material, an independent magnetization state of each magnet material can be achieved, as shown schematically in Figure 11.6(a) for a set of three independent micromagnetic elements. Here, a set of three magnetic actuators made from iron, NdFeB, and alnico are shown, and the magnetization direction of each actuator can be selectively switched by applying small or large magnetic fields.

As a second actuation scheme, a pair of magnetic materials can work together in one actuator, forming a magnetic composite whose magnetic moment sum interacts with externally applied or locally induced fields. Experimentally, we introduce a microscale permanent magnet composite material that can be remotely and reversibly turned off and on by the application of a magnetic field pulsed along the magnetic axis, which reverses the magnetization of one of the materials. For a completely remote operation, this pulsed field is supplied by electromagnetic coils outside the device workspace. This scheme is similar to *electropermanent magnets*, in which electromagnetic coils are wrapped directly around some of an array of switchable permanent magnets. When a strong current is pulsed through the coils, the magnetization of some of the permanent magnets is flipped, allowing for an off-on net magnetization of the set. Electropermanent magnets were originally used as centimeter-scale or larger magnetic work holders as an alternative to a mechanical vice [386]. While milliscale electropermanent magnets have been fabricated [387],

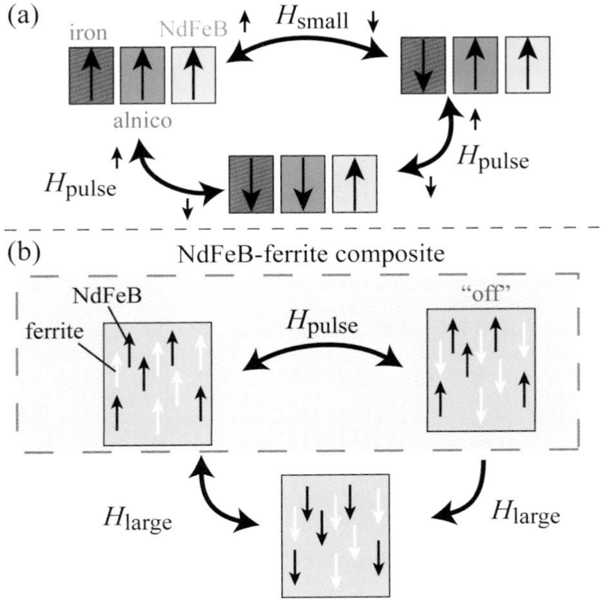

Figure 11.6

Schematic showing the multiple magnetic states, which can be achieved through the use of a variety of magnetic materials [57]. Copyright © 2012 by IEEE. Reprinted by permission of IEEE. (a) Three separate magnetic actuators, each made from a different magnetic material, the magnetization of which can be independently addressed by applying magnetic field pulses of various strengths. Here, H_{pulse} is a large field pulse and H_{small} is a small static field. (b) A single magnetic composite actuator can be switched between the "up," "off," or "down" states by applying pulses of different strength, where H_{large} is a large field pulse.

they contain integrated switching coils, preventing their scaling down to the micrometer scale for untethered operation.

The magnetic composite material presented here can be scaled down to the micron scale and enables remote wireless control. The anisotropic composite is made from two materials of equal magnetic moment: one permanent magnet material of high coercivity and one material, which switches magnetization direction by applied fields. By switching the second material's magnetization direction, the two magnets either work together or cancel each other, resulting in distinct on and off behavior of the device. The device can be switched on or off remotely using a field pulse of short duration. Because the switching field

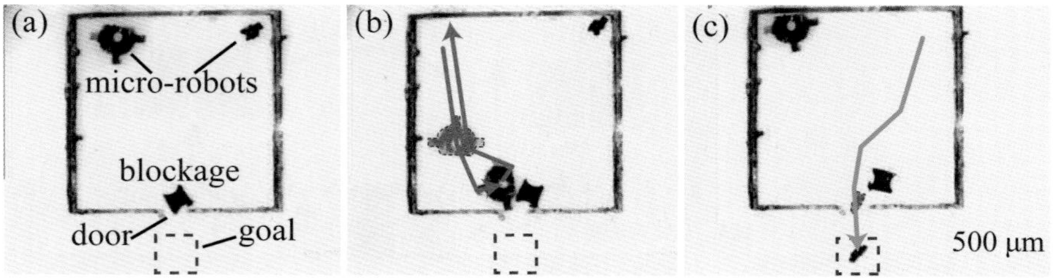

Figure 11.7

Addressable microrobot teamwork task, requiring the cooperative contribution of two mobile microrobots of different sizes working together to reach a goal [57]. Copyright © 2012 by IEEE. Reprinted by permission of IEEE. Frames show two superimposed frames, with the microrobot paths traced and midpoints outlined. (a) Both microrobots lie inside an enclosed area. The door to the goal is blocked by a plastic blockage. Only the larger microrobot can move the blockage, while only the smaller microrobot is small enough to fit through the door. (b) The larger microrobot is enabled and moved to remove the blockage while the smaller disabled microrobot remains in place. The larger microrobot is returned to its staring point and disabled. (c) The smaller microrobot is enabled and is free to move through the door to the goal.

pulse covers the entire workspace, this method could be used to selectively disable and enable many microdevices concurrently based on their orientations. Orientation control is achieved by a multi-step process using a field gradient to select a device for disabling by controlling each device's orientation.

To demonstrate the usefulness of a team of microrobots, a simple cooperative teamwork task is shown in Figure 11.7, where two microrobots of different sizes attempt to reach a goal location. Here, the two microrobots begin trapped in an enclosed area. The arena walls are made from polyurethane molded in a replica molding process similar to that used to fabricate the molded microrobots. The door to the goal is covered by a plastic blockage. Because the large microrobot is too big to fit through the door and the small microrobot is too small to move the blockage, both must work together as a team to reach the goal.

11.4 Homework

1. For a flexible active catheter device that is actuated magnetically inside an MRI system for cardiac applications, search the literature and list the possible localization techniques that could be used to track the catheter's position and motion with their pros and cons.

2. Discuss which multi-robot addressing technique described in Section 11.3 is the most scalable one (in the sense of maximum number of addressable microrobots) and can be used in magnetic microrobots moving in 3D.
3. Search the literature for microrobotic or actively propelled microparticle swarms. Report how they control such swarms. Discuss whether it is required to address each individual microrobot in such swarms for given potential applications.

12 Microrobot Applications

As it advances, microrobot technology has begun to be used in practical applications. Here, we overview the potential applications for mobile microrobots in the near future. Some progress has already been made in initial studies for many of these application areas, but much work needs to be done before the technology truly proves itself in each application area.

12.1 Micropart Manipulation

Manipulation could be used at the microscale to assemble parts or deliver payloads to goal locations. Such manipulation at the microscale requires precise actuation and control over adhesion forces to release manipulated parts [388–390]. This has traditionally made microscale manipulation challenging using microgrippers controlled by large robotic arms. Microrobots could offer advantages over these systems by providing remote manipulation inside enclosed spaces and could solve adhesion problems through liquid-based manipulation.

Methods for manipulating microscale objects can be classified into two categories: contact and non-contact manipulation. The distinction between the two is based on the presence or absence of physical contact when manipulating microparts. In general, contact manipulation is preferred for the study of microobjects that will not be damaged by any resulting contact forces. Contact manipulation can also supply larger pushing forces and increased speed. Non-contact methods are employed when manipulation forces must be comparatively low, if fine precision is required, or if the micropart is too fragile to be grasped by physical contact.

12.1.1 Contact-based mechanical pushing manipulation

Manipulation by direct mechanical contact can accomplished down to the micron scale by using "traditional" manipulation techniques using grippers fabricated using MEMS techniques. Such grippers are typically tethered, but one example [324] has made an untethered MEMS thermal gripper actuated by a focused laser. This design has been integrated into a levitating millirobot, which has three translational degrees of freedom over a small working space in an air environment. The manipulator is able to grasp and move objects from 100 μm to 1 mm in size for simple assembly tasks. However, as with all microgrippers operating in an air environment, the release of parts is a critical problem.

Remotely actuated manipulators must provide the precision and strength of traditional manipulators to be effective. While progress has been made, this is the major challenge in this application area.

In [35], magnetically levitated millirobots, which are moved by integrated electrical coil traces under the operating surface, are used to assemble simple centimeter-scale structures from a "bin" of parts. Solid parts are picked up using passive arms, and liquids such as glue are placed using a simple dipping arm. High speed and excellent potential for large-scale distributed manipulation are achieved with this system, with sub-micron precision when returning to a patterned trace location.

At the sub-millimeter scale, object manipulation becomes more difficult because controlled motion is difficult, and adhesive forces begin to overwhelm actuation forces. Therefore, all manipulation by sub-millimeter untethered microrobots has been conducted in a liquid environment to provide fluid damping and greatly reduced adhesive forces.

Simple magnetic microrobots are used to directly push microbeads [115] and cells [33] of sizes down to several microns. With controllable locomotion as slow as several microns per second, precise manipulation is possible by direct pushing using a relatively simple magnetic actuation. No specialized grippers are required for this manipulation.

The use of helical microswimmers offers 3D motion capability in liquid environments. In [45], microhelices are fabricated using 3D direct laser writing and vapor deposition and consist of a helical tail and a cage-like head to trap microparticles. Driving the helix toward a 6-μm colloidal particle resting on the substrate surface results in the particle being trapped in the cage, such that the helix can carry the particle in 3D to a goal. The particle is released by driving the helix in the reverse direction.

Preliminary demonstrations of manipulation using bio-hybrid microrobots has been shown [72], but more work must be done to steer such cargo for transport and delivery.

12.1.2 Capillary forces-based contact manipulation

In addition to contact mechanical manipulation methods, such as pushing, capillary forces can be used to pick and place microparts in air or inside fluids by controlling the contact angle of a fluid microdroplet in air or a microbubble inside fluids. For example, inside fluids, microbubbles can be trapped on the faces of a magnetic microrobot (see the robot SEM picture in Figure 4.1(b))

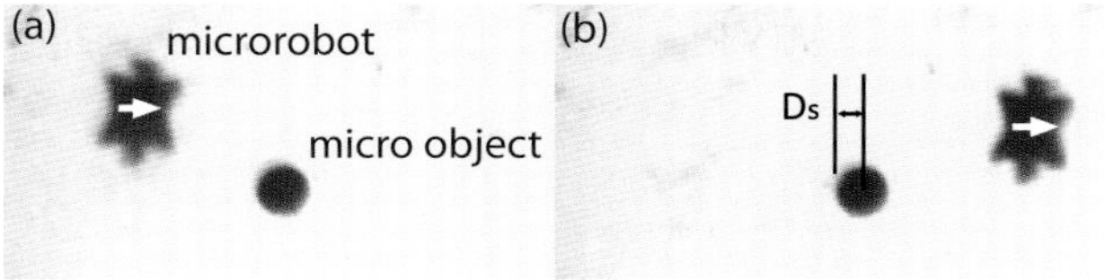

Figure 12.1

(a) A teleoperated star-shaped microrobot and a 210-μm microsphere for side-pushing under liquid on a glass surface [58]. Copyright © 2012 by IEEE. Reprinted by permission of IEEE. (b) The microrobot moves past the microsphere from its side, causing the sphere to displace a small amount of D_S, primarily due to the fluid interactions. Arrow on microrobot indicates direction of its motion.

to pick a variety of microparts, including biological tissues using attractive capillary forces [162, 391]. After transporting the parts via magnetic gradient pulling locomotion in 3D, external pressure has been used to decrease the microbubble's contact angle on the robot's face to release the parts easily.

12.1.3 Non-contact fluidic manipulation

It is also possible to manipulate objects in a low-Re fluid environment by using non-contact fluidic manipulation. In [115], a microrobot several hundred microns in size is used to manipulate microparts in a precise manner using fluid-based forces. The microrobot operates in a water environment, and translates past a micropart to exert forces. The motion of parts in the induced flow is a balance between the fluid drag force and the friction and adhesion, which acts to hold the particle still. The fluid boundary layer around the translating microrobot was studied in detail, with regions of influence defined for 1% of the maximum flow velocity. Using a 250-μm magnetic microrobot, 50- and 230-μm polystyrene spheres were manipulated without contact in this method. We now give a more detailed case study of this non-contact manipulation analysis for a single case of microrobot translation.

Case study: Non-contact manipulation using magnetic microrobots: Non-contact manipulation of microparts is possible using a translating magnetic microrobot near a microobject [58], as shown in Figure 12.1. For this analysis, we ignore any contact manipulation forces, and focus solely on the fluidic and surface adhesive forces. The surface forces between the microobject and the surface are taken from Section 3.4. Due to the choice of materials (polystyrene microspheres on glass in a water environment), the surface adhesion is negative, implying that its effects can be neglected. Viscous fluid drag

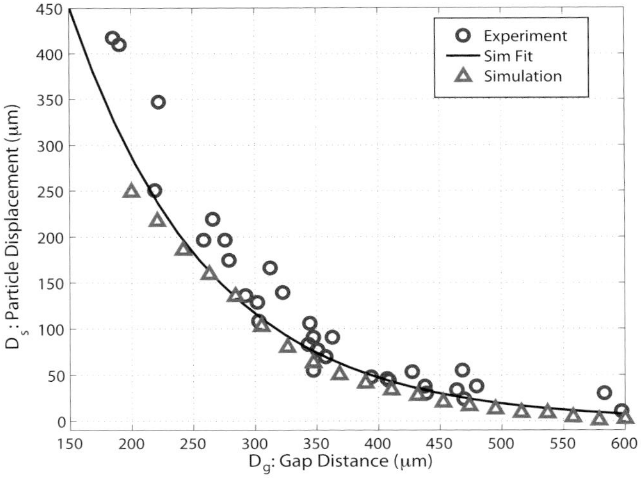

Figure 12.2

Simulation and experiment of a star-shaped microrobot manipulating a 210-μm microsphere from the side [58]. Copyright © 2012 by IEEE. Reprinted by permission of IEEE. Vertical division indicates whether sphere contact occurs with the microrobot's edge, determined from the simulation. The simulation fit "Sim Fit" is from the dynamic simulation, while "Sim Fit Lin" is a linear approximation to this fit, which can be used for control using these results.

is analyzed, as shown in Section 3.7.1. Because these objects are operating in a low-Re regime, the inertial effects of the microspheres can be neglected.

The fluid motion induced by the translating microrobot is attained by finite element modeling (FEM) using COMSOL Multiphysics (COMSOL Inc.). The low-Re (Stokes flow) physics is used, and the fluid velocity is found in the workspace. The microrobot is here modeled as stationary at an angle of $\pi/8$ radians with respect to the surface (an approximate average angle of the microrobot during its stick slip locomotion), and a bounding box defines the finite element simulation volume. The front and rear bounding faces are treated as a flow inlet and outlet, respectively, with a flow of 0.4 mm/s.

The fluid velocity due to the translating microrobot is shown in Figure 12.3, and the simulated particle motion found using a Runge-Kutta solver (ODE23s in MATLAB, Mathworks, Inc.).

This case study has shown the key physical parameters, which dictate the motion of microparticles in a robot-induced fluid flow. The fluid flow from a moving microrobot can be used to manipulate particles within microrobot body

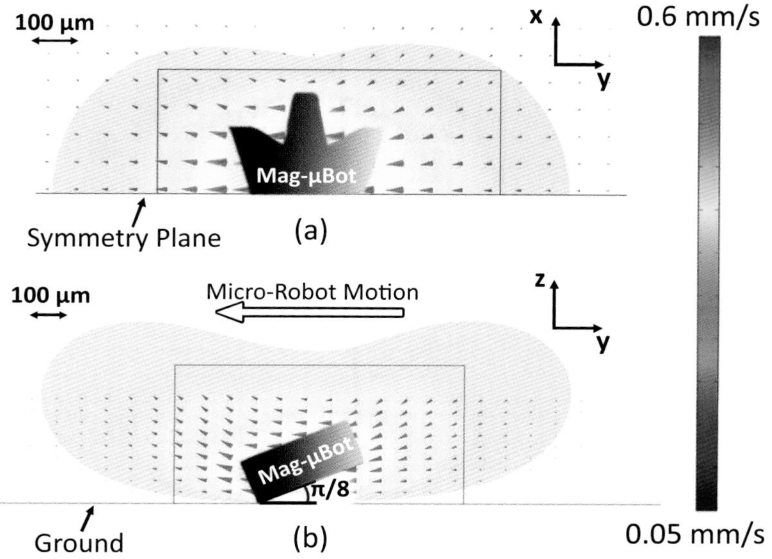

Figure 12.3

A side-view slice of the finite element modeling (FEM) solution for the flow around a star-shaped microrobot as it traverses through the environment [58]. Copyright © 2012 by IEEE. Reprinted by permission of IEEE. The microrobot is moving toward the left in these images, and the flow velocities correspond to *y*-directed flow, depicted by arrows. Half the microrobot is modeled in this analysis.

lengths away at slow speeds. This can be used for precise object positioning. As we have seen, to analyze this problem requires a full model of the microrobot motion, fluid flow from finite-element solutions, and microobject adhesion and friction model.

Rotational non-contact manipulation: As a related method in [59, 316], a constantly spinning micromanipulator was used to induce a rotational fluid flow, which moves microobjects in the region. Using a 380-μm microrobot, 200-μm particles are able to be moved at speeds up to 3.5 mm/s in a rotation motion. In addition, by using teams of these microrobots arranged in a reconfigurable grid pattern in 2D, complex "virtual channels" can be created allow the microrobots to pass the object along for long-distance fast transport, as shown in Figure 12.4. By tilting the microrobot rotation axis slightly from the vertical, rolling is achieved, which allows for precise microrobot positioning during manipulation. It is observed that a certain size of particle will become

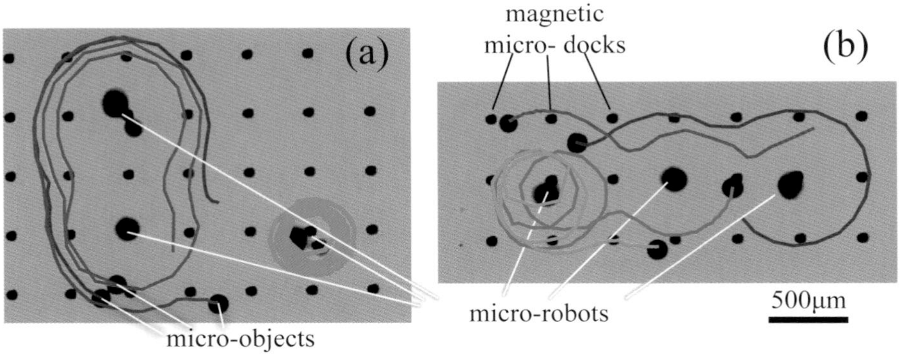

Figure 12.4

Team non-contact manipulation by three microrobots simultaneously spinning [59]. Copyright © 2012 by IEEE. Reprinted by permission of IEEE. The microrobot positions are trapped at discrete locations by magnetic docks embedded in the surface. Manipulated microsphere paths are tracked by colored lines.

trapped near the microrobot for certain rotation speeds, and so the microrobot can carry the particle over long distances using this method. By combining these long-distance and slower precision rotational flow manipulation methods, a course-fine object placement is demonstrated.

Rotational flows can create a trap around a spinning and rolling spherical magnetic microrobot. Trapped microobjects can also be transported by the rolling locomotion of the robot. As an example demonstration, live or motile cells such as bacteria [60] are trapped and transported, as shown in Figure 12.5. It is possible to selectively trap and transport individual freely swimming multi-flagellated bacteria over a distance of 30 μm (7.5 body lengths of the carrier) on a surface, using the rotational flows locally induced by the rotating microrobot. Only a weak uniform magnetic field (<3 mT) is needed to rotate the robot. The microrobot can translate on a substrate by rotating at a speed of up to 100 μm/s while providing a fluidic trapping force of a few to tens of pico-Newtons.

At a smaller scale, this was also executed using a rotating nanowire or self-assembled collection of microbeads [392, 393]. Using a weak rotating magnetic field, this approach was used to move microspheres and cells using a 13-μm long nickel nanowire. Thus, such non-contact manipulation methods have shown their effectiveness across several size scales.

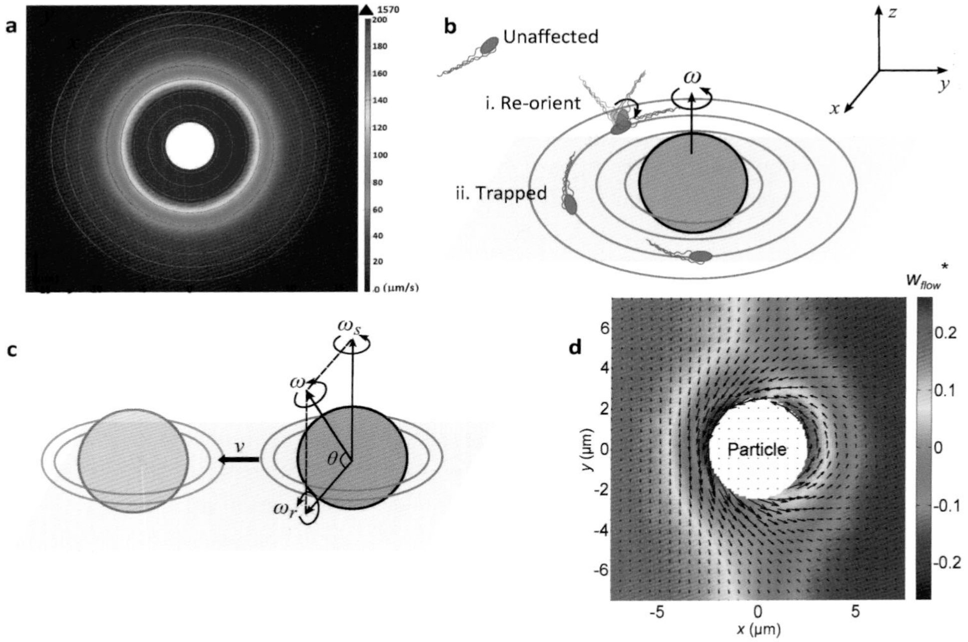

Figure 12.5

Trapping and translating live cells or other microentitites by a spinning and rolling magnetic spherical microrobot [60]. Reproduced with the permission of the Royal Society of Chemistry. (a) The simulation result of a 5-μm diameter spherical microrobot spinning at 100 Hz on a flat surface in water. The plot is from a top view of the cross-section taken at the equatorial plane of the microrobot. Red concentric circles represent the streamlines. The color map shows the flow velocity distribution. (b) The schematic of the trapping of a nearby bacterium by the rotational flow induced by the rotation ω of the robot near a flat surface. Any bacterium that is far away is minimally affected by the induced flow. A bacterium that is close enough to the spinning particle is first reoriented by the flow to align its body's long axis with the local streamline (i). Then it is trapped (ii) and orbited around the particle. (c) The schematic of the mechanism for enabling the mobility of the induced rotational flow field. Instead of being perpendicular to the surface as shown in (b), the rotation axis of the particle is tilted from the z-axis. (d) The finite element simulation result of a 5-μm diameter spherical microrobot rotating at 100 Hz with a tilt angle of 75 degrees and a translational speed of $0.06\omega_r a$ in the $-y$ direction on a flat surface in water. The plot is from a top view of the cross-section taken at the equatorial plane of the robot. The arrows indicate the in-plane flow velocity at selective positions, while the color map shows the distribution of the out-of-plane flow velocity normalized by the magnitude of the in-plane flow velocity at the same position.

Artificial bacterial flagella can also be used for non-contact object manipulation. The rotational motion of these microrobots naturally creates a rotational fluid flow, which can be used to move objects [394]. The coordinated manipulation of many microobjects has been shown in this way, for example, to clear an area of particles, although it is not clear whether this can be used for precision manipulation of particles to goal positions as the fluid flow is necessarily coupled to the microhelix translation.

12.1.4 Autonomous manipulation

Almost all current microrobotic manipulation systems are teleoperated by a human user, which limits the precision, repeatability, and speed of the manipulation tasks. Most microassembly tasks with multiple microobjects takes minutes or hours due to such slow speed. Therefore, autonomous manipulation control methods using visual and other sensory feedback are crucial for the future practical manipulation and other applications of microrobots.

As the only example of autonomous manipulation study using mobile microrobots so far, contact and non-contact micromanipulation methods have been used by a surface crawling microrobot for precision assembly of two microspheres in 2D autonomously [58]. The robot path planning has been conducted by the Wavefront algorithm using visual feedback. One difficulty with precision manipulation using contact methods is that retraction motion of the microrobot after the assembly of two target objects moves the assembled objects also. Using principles of non-contact manipulation from [115], an autonomous particle manipulation controller has been developed, which uses physical models and an iterative learning controller to precisely manipulate particles using non-contant fluid forces. Even in the presence of unknown disturbance forces, the model-based feed-forward input of this controller allows for precise manipulation and retraction of the microrobot afterward. In addition, this work presented the assembly of two particles together, a task that is generally difficult using non-contact manipulation.

12.1.5 Bio-object manipulation

The manipulation of bio-objects by untethered microrobots has much promise for lab-on-a-chip applications, individual cell study, and tissue scaffolds. A major requirement for such manipulation is gentle pushing so as not to damage the object, as well as bio-compatibility. In [277], a bubble microrobot is

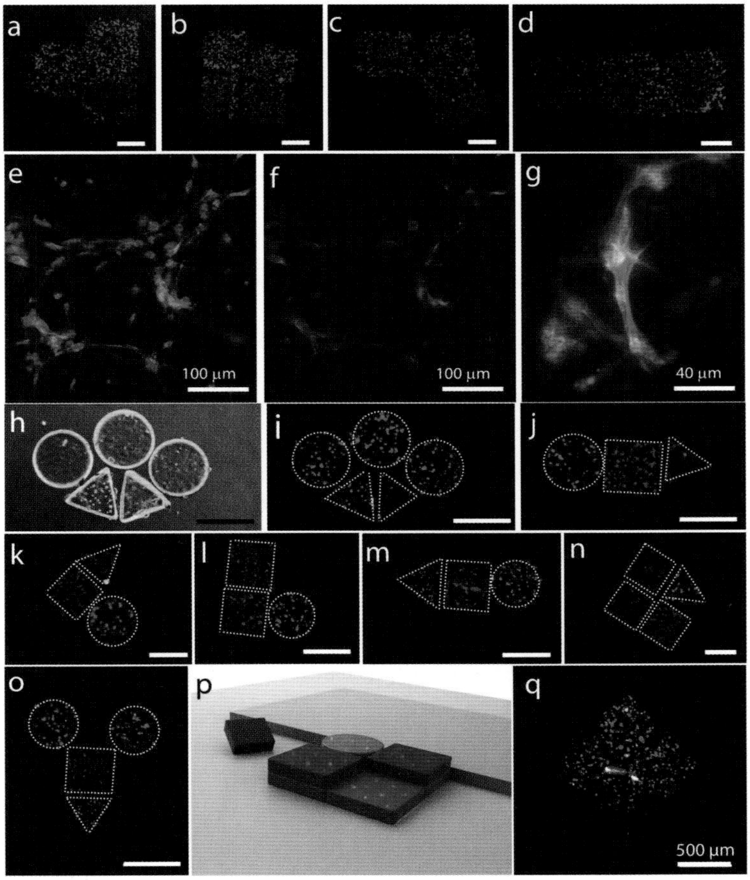

Figure 12.6

Microgel blocks with embedded live cells have been assembled using a magnetically driven microrobot on a planar surface in physiological fluids [27]. Copyright © 2014 by Nature Publishing Group. Reprinted by permission of Nature Publishing Group. Fluorescence images of NIH 3T3 cell-encapsulating hydrogels after the assembly of (a) T-shape, (b) square-shape, (c) L-shape, and (d) rod-shape constructs. Green represents live cells and red represents dead cells. (e-g) Immunocytochemistry of proliferating cells stained with Ki67 (red), DAPI (blue), and Phalloidin (green) at day 4. (e) Cells stained with DAPI and Phalloidin at 20× magnification. (f) Cells stained with Ki67 and Phalloidin at 20× magnification. (g) Cells stained with Ki67, DAPI, and Phalloidin at 40× magnification. (h-q) 2D and 3D heterogeneous assemblies of HUVEC, 3T3, and cardiomyocyte-encapsulating hydrogels. HUVECs, 3T3s, and cardiomyocytes are stained with Alexa 488 (green), DAPI (blue), and Propidium iodide (red), respectively. (h) Bright field and (i) fluorescence images of an assembly composed of circular and triangular gels. (j-o) Fluorescence images of several 2D heterogeneous assemblies of HUVEC, 3T3, and cardiomyocyte-encapsulating hydrogels. (p) Schematic form and (q) fluorescence image of 3D heterogeneous assembly of HUVEC, 3T3, and cardiomyocyte-encapsulating hydrogels. Scale bars are 500 µm unless otherwise is stated.

shown to manipulate a number of hydrogels functionalized with yeast cells. The gels are arranged into a tight heterogeneous 2D grid by the microrobots, and after time the yeast cells are cultured on the scaffold. Such a demonstration has potential use for complex cell culture experiments or for in-vitro growth of tissues or even organs. In [27], microgels embedded with live cells have been assembled using a magnetically driven microrobot on a planar surface in physiological fluids in 2D and 3D, as shown in Figure 12.6. Even, an untethered magnetic microgripper can be used to pick and place such microgels mechanically [26, 395] or using capillary forces on microbubbles [162] to enable 3D microassembly in liquid media.

12.1.6 Team manipulation

Manipulation of microobjects by teams of microrobots could have major advantages in speed and capability. In [396], multiple magnetic microrobots are moved independently using an electrostatic trapping surface. Because every untrapped microrobot moves in parallel, this approach requires careful planning to increase manipulation speed over a single-robot case.

Using teams of microrobots trapped at "docking" sites in the substrate, rotating magnetic microrobot teams have been used to perform non-contact manipulation of objects in a liquid environment [59]. Because the location of the spinning microrobots is variable, the microrobots in this case form "virtual channels", which move objects in a versatile manner appropriate for use in microfluidic channels.

As a method that naturally allows for multi-robot control, optically controlled bubble microrobots have been used to perform team manipulation of objects [277]. In this work, two microrobots are used to sandwich microobjects for precise and fast manipulation.

These team demonstrations are promising for distributed and parallel manipulation, but must prove their advantage over the much simpler single-robot manipulation case for it to adopted.

12.1.7 Microfactories

Microrobots working in 2D or 3D could be used to assemble microparts in ways that are difficult or not possible using conventional fabrication techniques. Of particular interest is the assembly of 3D parts, which require orientation and position control. Microrobots could apply adhesives, position parts, and repair defects in a desktop setup. Because such a process would likely be a serial assembly process, it could benefit greatly from parallel microrobot assembly teams. While micropart manipulation thus far has not approached the sophistication required for such a microfactory, the potential is great, and the concept has been proven at a slightly larger size scale. Pelrine et al. [35] have shown such a process with magnetic millirobots levitating on a diamagnetic surface. Each robot in this study was equipped with a tool such as a gripper, adhesive applicator, or weighing pan for distributed operation.

12.2 Health Care

Remote microrobots have great promise for medical applications [55, 67, 251]. Some of the potential application areas for medical microrobots inside the human body are thoroughly outlined in [397] and [55] with their potential opportunities and challenges. Such application areas can be listed as:

- Cargo (drug, gene, RNA, stem cell, etc.) delivery in targeted local regions in controlled amounts,
- Brachytherapy,
- Marking target therapy areas,
- In situ monitoring of chemical concentrations, pressure, pH, temperature, etc.,
- Electrode implantation,
- Creating or opening occlusions,
- Tissue scaffold creation,
- Biopsy sampling [398],
- Thermal or mechanical ablation,
- Cauterization, and
- Hyperthermia treatment.

Some of the targeted areas of the body could include the circulatory system, central nervous system, gastrointestinal tract [399–401], urinary tract, eye, and auditory canal. A recent review covers potential and recent use of bacteria-based microswimmers or bacteria directly for targeted drug or other cargo delivery applications [402].

As first steps towards these application areas, a magnetically controlled needle is shown to operate inside artificial and ex-vivo eyes [16]. As a health-care application area, the inside of the eye is a natural first step as the volume of the eye is visible through an optical microscope. This preliminary therapy aims to puncture the vasculature. Navigation inside the eye is, however, complicated due to the complex optics of the eye and non-Newtonian fluid inside the eye. An algorithm to compensate for the optical distortion is presented [376], which obtains the 3D position of an intraocular microrobot, assuming the microrobot geometry is known. Drug delivery inside the eye is investigated with a drug that is coated on a microrobot surface [403], where the drug is diffused near to the target region for potential treatment of retinal vein occlusion.

Progress in the other potential medical application areas will come with refinement of 3D microrobot locomotion strategies in non-Newtonian biological fluids and development of functional microrobots with integrated microtools or sensing, cargo carrying, heating, and other functional capabilities.

12.3 Environmental Remediation

Self-propelled microrobots can be used for future environmental remediation technology [404]. Microrobot-enabled degradation and removal of major contaminants and microrobot-based water quality monitoring are example environmental applications of microrobots. Future autonomous microrobot swarms could monitor and respond to hazardous chemicals and use chemotactic and pH-tactic search strategies to trace chemical plumes to their source. As an example, usable Fe/Pt multi-functional active microcleaners are proposed for waste-water treatment and water reuse, which is an essential part of environmental sustainability [405]. These microcleaners are capable of degrading organic pollutants (malachite green and 4-nitrophenol) by generated hydroxyl radicals via a Fenton-like reaction. Such microcleaners can continuously swim for more than 24 hours and can be stored more than 5 weeks during multiple cleaning cycles. They can be also reusable, which can reduce the cost of the process.

12.4 Reconfigurable Microrobots

The field of reconfigurable robotics proposes versatile robots that can reform into various configurations depending on the task at hand [406]. These types of robotic systems consist of many independent and often identical modules, each capable of motion and combining with other modules to create assemblies. Then these modules can be disassembled and reassembled into alternate configurations. For example, SuperBot consists of 20 modules that can combine to form a mobile mechanism that can roll across the ground for 1 km and then reconfigure into one that can climb obstacles [407].

Another concept in the field of reconfigurable robotics is *programmable matter*, which is an active matter that can assemble and reconfigure into programmed 3D shapes, giving rise to synthetic reality [408]. This is similar to virtual or augmented reality, where a computer can generate and modify an arbitrary object. However, in synthetic reality, this object has physical realization. A primary goal for programmable matter is scaling down the size of each individual module, with the aim of increasing spatial resolution of the final assembled product. Currently, the smallest deterministic, actuated module in a reconfigurable robotic system fits inside a 2 cm cube [409], which is a self-contained module that is actuated using shape memory alloy. Scaling down further into the sub-millimeter scale brings new issues, including module fabrication, control, and communication.

For the purposes of microscale assembly using microrobots, Donald et al. [38] demonstrate the assembly of four microfabricated silicon microrobots, each under 300 μm in all dimensions, actuated by electric fields. Once assembled however, they cannot detach and reconfigure, because the electrostatic driving fields do not allow for disassembly. Lipson et al. have demonstrated reconfigurable assemblies using 500-μm planar silicon elements [410] and centimeter-scale 3D elements [411]. By controlling the local fluid flow in these systems, the elements can be deterministically assembled and disassembled into target shapes. This system relies on an active substrate to provide fluid flow and control and so the assembled microscale building blocks have limited mobility.

In [61], sub-millimeter scale untethered permanent magnet microrobots (Mag-μBots) are actuated by external magnetic fields as components of magnetic micromodules (Mag-μMods) for creating deterministic reconfigurable 2D microassemblies; this implies that the Mag-μMods will be able to both assemble and disassemble. Strong permanent magnet modules will attract each other with large magnetic forces; therefore it is necessary to reduce this magnet force between modules to facilitate disassembly. This can be achieved by adding an outer shell to the Mag-μBot for the design of a module. The outer shell prevents two magnetic modules from coming into close contact, where magnetic forces will become restrictively high. However, they are still sufficiently close together to yield a mechanically stable assembly.

A critical design and control component of reconfigurable microrobots is the reversible attachment and detachment method for each module. Magnetic forces become strong after the magnetic modules attach each other, and detachment needs extra surface clamping and torque actuation. As other alternative reversible bonding methods at the micron scale, gecko foot-hairs-inspired reversible elastomer microfiber adhesives [137, 412–436], heat-activated thermoplastic bonding materials [437], heat-activated liquid metal bonding materials [438], and other light- and heat-stimulated switchable adhesive methods can be used. In microfiber adhesives, modules need to be pressed gently to each other to attach/adhere, and they need to be rotated by twisting to mechanically peel off. As the main advantages of such microfiber adhesives are not requiring any external stimulus, such as heat or light to active or deactivate the bonding (it is pure mechanical contact loading and peeling enabling attachment and detachment, respectively), being highly repeatable and reversible, and being scalable down to tens of microscale modules by using nanoscale fiber adhesives. Such elastomer fibers mainly use surface forces such as van der Waals forces to adhere surfaces after a close contact [412]. In [437], nine magnetic microrobot modules have been bonded by a thermoplastic layer around the modules. However, detachment of the modules was challenging because thermoplastic material left residue and was not repeatable. On the other hand, phase changing liquid Gallium-based bonding method of a microrobot with other microobjects [438] was repeatable with no residue left on the other surface, and such liquid metal material needed only 7 to 10° temperature increase from the room temperature to change the phase of Gallium layer from solid to liquid. Therefore, it is a promising reversible bonding method for reconfigurable micromodules with remote or on-board heating capability.

Motion of multiple Mag-μMods is achieved by employing a surface divided into a grid of cells, where each cell on the surface contains an addressable electrostatic trap capable of anchoring individual Mag-μMods to the surface by capacitive coupling; this prevents them from being actuated by the external magnetic fields. This approach is related to that found in the field of *distributed manipulation* [377], where parts are manipulated in parallel using programmable force fields, but here the distributed cells provide only a retarding force while the actuation magnetic force is globally applied to all modules. Unanchored Mag-μMods can move on the surface due to the imposed

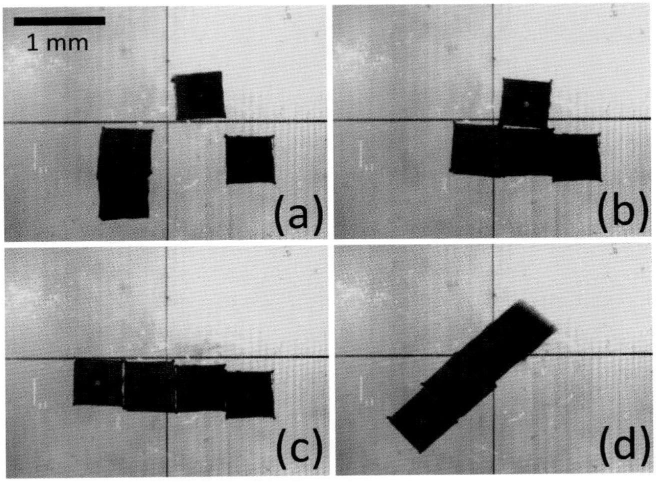

Figure 12.7

Frames from a movie with four teleoperated Mag-μMods assembling into a reconfigurable structure [61]. Copyright © 2011 by SAGE Publications, Ltd. Reprinted by permission of SAGE Publications, Ltd. Arrows indicate direction of magnetization. (a) Four Mag-μMods prepare for assembly. (b) All four modules are assembled in a T-configuration. (c) One module is broken free by rotation and reattaches in a new configuration. (d) The new assembly is mobile, and is shown moving to a new location.

magnetic fields, and move in parallel. This technique is identical to controlling multiple Mag-μBots, explained in detail in [56]. Assembling two Mag-μMods is straightforward – by moving an non-anchored Mag-μMod towards an anchored one, magnetic forces eventually dominate and cause the two Mag-μMods to self-assemble.

For disassembly of two Mag-μMods to occur, the magnetic attraction between them must be overcome to separate them. To do this, the electrostatic grid surface is used to anchor parts of assembled modules, and examine the effectiveness of externally applied magnetic torques to disassemble non-anchored modules from the assembly.

Figure 12.7 shows the concept of multiple Mag-μMods assembling, disassembling, and reconfiguring into a different configuration. Because the Mag-μMods are magnetic, they can only assemble into configurations that are magnetically stable, implying that they can form single closed flux loops.

12.5 Scientific Tools

Untethered microrobots are capable of exerting forces on objects potentially inaccessible by other methods. Such action can be used to probe living organisms or microscale structures as a diagnostic investigation of material properties and mechanical response. While sophisticated investigations have begun with tethered microrobotic setups [439, 440], the use of untethered microrobots has potential to perform these studies in natural environments. The direct manipulation of cells can also be important for study in microfluidic channels. It has been demonstrated that a magnetic microrobot can push live cells without damaging the cells [33] for localized observation and probing.

The physical probing of individual cells to investigate mechanobiological response has been studied [441]. In this work, a magnetic microtool (MMT) is integrated with an on-board force sensor based on visual detection of beam deflection. The MMT has a sharp tip and is used in a preliminary work to mechanically stimulate a 100-μm diatom cell while the response is observed.

These preliminary studies have shown that microrobots can form tools for study of phenomena at the microscale. It is expected that similar applications in materials research, bio-technology, microfluidics, and other areas will be possible with appropriately designed microrobots with the ability to move object, apply precise forces, measure chemical concentrations, and other abilities.

Remote sensing using a mobile microrobot could be used as a tool to investigate and map chemical concentrations, temperature, and so on in enclosed spaces in high resolution. Optically detected oxygen sensors have been integrated with mobile microrobots capable of 3D motion for interrogation inside the human eye [181]. Additional modalities, such as resonant-based sensors read by reflected light [442] or magnetic fields [443], could increase the versatility of these remote measurements.

13 Summary and Open Challenges

13.1 Status Summary

The field of microrobotics has proven itself to be an exciting potential solution to problems in confined microscale spaces. Research thus far has focused on extending robotics principles of precise top-down control to the sub-millimeter-size scale for motion and interaction with the environment. Many novel solutions have been designed to operate effectively in the presence of microscale physics, in particular with high surface adhesion and viscous fluid flow. Current demonstrations in microrobotics have primarily focused on methods of moving in 2D or 3D. Solutions in 2D have included crawling on surfaces in air or liquid environments using a wide variety of actuation methods. The diversity of the approaches is promising, as each method contains particular advantages and disadvantages which could align with potential application areas. The demonstrated methods are capable of precisely moving microrobots using traditional robotics techniques such as path planning and obstacle avoidance, and have been shown in remote inaccessible spaces such as inside microfluidic channels. In 3D, motion has been accomplished by swimming or pulling by chemical reactions or magnetic forces. Some of these methods have been biologically-inspired, using novel swimming mechanisms from microorganisms to move in viscous low-Re fluid environments. These methods have allowed for feedback-control path following in 3D and have been proven to work in difficult environments, such as in the presence of fluid flow.

Feedback of microrobot position has thus far primarily taken the form of vision through a microscope. This has provided precise localization in 2D using a single camera or in 3D using multiple cameras or advanced vision algorithms, which is naturally suited for machine vision feedback control. While alternate localization by ultrasound, X-ray, and MRI have been investigated, these techniques are still in the proof-of-concept stage for microrobotics.

Some progress has been made towards microrobot applications in object manipulation, in healthcare, and as tools for scientific study. Thus far, these have only been proof-of-concept demonstrations, but such research has been increasing recently as more microrobot capabilities increase.

13.2 What Next?

Despite such progress, many open challenges remain in the field of microrobotics. Some of these have been identified in this book, and some are inevitably yet to be known at all due to the fast-changing nature of the field. The desired microrobot capabilities are primarily driven by the potential applications. Some of the areas where concrete advances are required are listed next:

- **Locomotion**. Significant progress must be made in microrobot locomotion in 2D and 3D. In particular, precision, speed, and high force capabilities must be improved to allow microrobots to be useful tools for microscale applications. Current methods are promising, but must be moved out of the concept stage and be developed into "technologies" which can be applied to solve other problems.
- **Multi-robot control**. While a few methods for multi-robot control have been demonstrated, these methods suffer from poor scalability to large numbers of microrobots. For parallel distributed operation to show its full potential, large numbers of microrobots must operate in tandem in 2D or 3D. Whether this takes the form of greater numbers of individually-addressed microrobots or swarm-like interactions between agents resulting in emergent team behaviors, this challenging research area could have a large impact on the usefulness of microrobots over existing technologies.
- **Localization**. Optical microscope-based localization is adequate for a few microrobotics applications, such as in microfactories or microfluidics, but is not compatible with many applications, most notably use inside the human body. Thus, alternate methods must be developed which provide precision, high feedback rates, and compatibility with medical procedures.
- **Part manipulation**. Manipulation procedures must be improved to allow for precise part transport and assembly. Methods for moving objects in 3D and over long distances must be designed to open up the design space for microassembly applications. Applications for micromanipulation by microrobots must target areas where existing microassembly by photolithography or other methods is inadequate. These could include creating complex 3D assemblies, or in providing a versatile and flexible assembly paradigm.
- **Tools**. Integrated tools must be designed to exploit the full potential of mobile microrobots. Methods for heating, gripper, cutting, or others which are wirelessly actuated along with the microrobot could transform microrobots from interesting novelties to truly useful devices for interacting with the environment.

- **Sensing**. Integrating sensing functionality to mobile microrobots is indispensable to enable many high-impact mobile sensor network applications in medicine, inspection, and environmental monitoring. Such mobile microsensors could probe microfluidic channels or the inside of the human body in a location-specific manner not possible by any other method. The sensing method must work remotely, with or without visual line of sight.
- **Communication**. Wireless communication among autonomous microrobots is an unsolved challenge, which needs to be resolved if a given application requires distributed agents with their own computing, powering, and communication capability. At the sub-millimeter length scale, current RF communication techniques are not practical to implement while optical communication might be possible. In addition, new communication methods using chemicals (similar to quorum sensing among bacteria), electrical or magnetic signals, vibrations, color change, and so on among microrobots need to be developed.
- **Biocompatibility and biodegradability**. For any biotechnology or medical applications, microrobots must be biocompatible. High-strength electric fields, toxic magnetic materials, or high temperatures are all potential problem areas for biocompatibility. This problem will be solved by a judicious choice of actuation method along with biocompatible coatings and packaging for toxic microrobot materials. Moreover, many medical microrobots would not be possible to be extracted from the human body after their operation is over, and they might cause health hazards if they remain inside the body. Therefore, the ideal solution would be to fabricate these robots from biodegradable materials so that they self-degrade in a given duration in the given environmental conditions.
- **Automation and learning**. As the complexity of microrobotic systems increase, more sophisticated control algorithms, including autonomous control, must be further developed. Most demonstrated microrobot tasks thus far have been controlled via teleoperation or simple path following algorithms. In addition, to accommodate changing experimental conditions and fabrication errors, adaptive learning control algorithms will be required for robust operation [58].
- **Self-organization, collectivity, and swarming**. Because it is mostly not possible or practical to control individual microrobots when they are in large numbers as a swarm, one of the efficient ways to control them is to design

and control local interactions among microrobots to self-organize them collectively. Such local interactions could be magnetic, fluidic, electrostatic, surface tension based, or based on other long- or short-range forces at the microscale, and they could be remotely controlled or tuned to enable different self-organized patterns at a given time.

- **Applications**. Real-world unique and high-impact applications of microrobots in medicine, environment, microfactories, biotechnology, and so on need to still be demonstrated. Current demonstrations are only proof of concept and preliminary toward clinical or industrial use. For example, medical microrobots achieving disease diagnosis, treatment, or surgery in currently non-accessible regions of the human body non-invasively need to be demonstrated by pre-clinical in vivo animal model experiments for specific diseases that have no other existing medical technology solution at the moment.

Logical approaches to some of these action areas have been identified, while the best approach to other areas is still an open question.

Bibliography

[1] D. C. Meeker, E. H. Maslen, R. C. Ritter, and F. M. Creighton, "Optimal realization of arbitrary forces in a magnetic stereotaxis system," *IEEE Transactions on Magnetics*, vol. 32, pp. 320–328, Nov. 1996.

[2] K. Ishiyama, M. Sendoh, A. Yamazaki, and K. Arai, "Swimming micro-machine driven by magnetic torque," *Sensors and Actuators A: Physical*, vol. 91, pp. 141–144, June 2001.

[3] N. Darnton, L. Turner, K. Breuer, and H. C. Berg, "Moving fluid with bacterial carpets," *Biophysical Journal*, vol. 86, pp. 1863–1870, Mar. 2004.

[4] W. F. Paxton, K. C. Kistler, C. C. Olmeda, A. Sen, S. K. St. Angelo, Y. Cao, T. E. Mallouk, P. E. Lammert, and V. H. Crespi, "Catalytic nanomotors: autonomous movement of striped nanorods," *Journal of the American Chemical Society*, vol. 126, no. 41, pp. 13424–13431, 2004.

[5] R. Dreyfus, J. Baudry, M. L. Roper, M. Fermigier, H. A. Stone, and J. Bibette, "Microscopic artificial swimmers," *Nature*, vol. 437, pp. 862–865, Oct. 2005.

[6] N. Mano and A. Heller, "Bioelectrochemical propulsion," *Journal of the American Chemical Society*, vol. 127, no. 33, pp. 11574–11575, 2005.

[7] S. Martel, J. B. J.-B. Mathieu, O. Felfoul, A. Chanu, E. Aboussouan, S. Tamaz, P. Pouponneau, G. Beaudoin, G. Soulez, M. Mankiewicz, and L. Yahia, "Automatic navigation of an untethered device in the artery of a living animal using a conventional clinical magnetic resonance imaging system," *Applied Physics Letters*, vol. 90, no. 11, p. 114105, 2007.

[8] B. R. Donald and C. G. Levey, "An untethered, electrostatic, globally controllable MEMS micro-robot," *Journal of Microelectromechanical Systems*, vol. 15, no. 1, pp. 1–15, 2006.

[9] O. Sul, M. Falvo, R. Taylor, S. Washburn, and R. Superfine, "Thermally actuated untethered impact-driven locomotive microdevices," *Applied Physics Letters*, vol. 89, p. 203512, 2006.

[10] S. Martel, J. B. Mathieu, O. Felfoul, A. Chanu, E. Aboussouan, S. Tamaz, P. Pouponneau, G. Beaudoin, G. Soulez, and M. Mankiewicz, "Automatic navigation of an untethered device in the artery of a living animal using a conventional clinical magnetic resonance imaging system," *Applied Physics Letters*, vol. 90, p. 114105, 2007.

[11] L. Zhang, J. Abbott, L. Dong, B. Kratochvil, D. Bell, and B. Nelson, "Artificial bacterial flagella: fabrication and magnetic control," *Applied Physics Letters*, vol. 94, p. 064107, 2009.

[12] A. Ghosh and P. Fischer, "Controlled propulsion of artificial magnetic nanostructured propellers," *Nano Letters*, vol. 9, pp. 2243–2245, June 2009.

[13] C. Pawashe, S. Floyd, and M. Sitti, "Modeling and experimental characterization of an untethered magnetic micro-robot," *International Journal of Robotics Research*, vol. 28, no. 8, pp. 1077–1094, 2009.

[14] A. A. Solovev, Y. Mei, E. Bermúdez Ureña, G. Huang, and O. G. Schmidt, "Catalytic microtubular jet engines self-propelled by accumulated gas bubbles," *Small*, vol. 5, pp. 1688–1692, July 2009.

[15] S. Martel and M. Mohammadi, "Using a swarm of self-propelled natural microrobots in the form of flagellated bacteria to perform complex micro-assembly tasks," in *International Conference on Robotics and Automation*, pp. 500–505, 2010.

[16] M. P. Kummer, J. J. Abbott, B. E. Kratochvil, R. Borer, A. Sengul, and B. J. Nelson, "OctoMag: An electromagnetic system for 5-DOF wireless micromanipulation," *IEEE Transactions on Robotics*, vol. 26, no. 6, pp. 1006–1017, 2010.

[17] H.-R. Jiang, N. Yoshinaga, and M. Sano, "Active motion of a janus particle by self-thermophoresis in a defocused laser beam," *Physical Review Letters*, vol. 105, no. 26, p. 268302, 2010.

[18] W. Hu, K. S. Ishii, and A. T. Ohta, "Micro-assembly using optically controlled bubble microrobots," *Applied Physics Letters*, vol. 99, no. 9, p. 094103, 2011.

[19] A. Buzas, L. Kelemen, A. Mathesz, L. Oroszi, G. Vizsnyiczai, T. Vicsek, and P. Ormos, "Light sailboats: Laser driven autonomous microrobots," *Applied Physics Letters*, vol. 101, no. 4, p. 041111, 2012.

[20] W. Wang, L. A. Castro, M. Hoyos, and T. E. Mallouk, "Autonomous motion of metallic microrods propelled by ultrasound," *ACS Nano*, vol. 6, no. 7, pp. 6122–6132, 2012.

[21] V. Magdanz, S. Sanchez, and O. Schmidt, "Development of a sperm-flagella driven micro-bio-robot," *Advanced Materials*, vol. 25, no. 45, pp. 6581–6588, 2013.

[22] J. Zhuang, R. W. Carlsen, and M. Sitti, "pH-taxis of bio-hybrid microsystems," *Scientific Reports*, vol. 5, p. 11403, 2015.

[23] R. W. Carlsen, M. R. Edwards, J. Zhuang, C. Pacoret, and M. Sitti, "Magnetic steering control of multi-cellular bio-hybrid microswimmers," *Lab on a Chip*, vol. 14, no. 19, pp. 3850–3859, 2014.

[24] D. Kim, A. Liu, E. Diller, and M. Sitti, "Chemotactic steering of bacteria propelled microbeads," *Biomedical Microdevices*, vol. 14, pp. 1009–1017, Sept. 2012.

[25] E. Diller, J. Zhuang, G. Z. Lum, M. R. Edwards, and M. Sitti, "Continuously distributed magnetization profile for millimeter-scale elastomeric undulatory swimming," *Applied Physics Letters*, vol. 104, no. 17, p. 174101, 2014.

[26] E. Diller and M. Sitti, "Three-dimensional programmable assembly by untethered magnetic robotic micro-grippers," *Advanced Functional Materials*, vol. 24, no. 28, pp. 4397–4404, 2014.

[27] S. Tasoglu, E. Diller, S. Guven, M. Sitti, and U. Demirci, "Untethered micro-robotic coding of three-dimensional material composition," *Nature Communications*, vol. 5, article no: 3124, 2014.

[28] K. K. Dey, X. Zhao, B. M. Tansi, W. J. Méndez-Ortiz, U. M. Córdova-Figueroa, R. Golestanian, and A. Sen, "Micromotors powered by enzyme catalysis," *Nano Letters*, vol. 15, no. 12, pp. 8311–8315, 2015.

[29] A. Servant, F. Qiu, M. Mazza, K. Kostarelos, and B. J. Nelson, "Controlled in vivo swimming of a swarm of bacteria-like microrobotic flagella," *Advanced Materials*, vol. 27, no. 19, pp. 2981–2988, 2015.

[30] W. Gao, R. Dong, S. Thamphiwatana, J. Li, W. Gao, L. Zhang, and J. Wang, "Artificial micromotors in the mouses stomach: A step toward in vivo use of synthetic motors," *ACS Nano*, vol. 9, no. 1, pp. 117–123, 2015.

[31] E. Diller, J. Giltinan, G. Z. Lum, Z. Ye, and M. Sitti, "Six-degree-of-freedom magnetic actuation for wireless microrobotics," *International Journal of Robotics Research*, vol. 35, no. 1-3, pp. 114–128, 2016.

[32] D. R. Frutiger, K. Vollmers, B. E. Kratochvil, and B. J. Nelson, "Small, fast, and under control: Wireless resonant magnetic micro-agents," *International Journal of Robotics Research*, vol. 29, pp. 613–636, Nov. 2009.

[33] M. S. Sakar and E. B. Steager, "Wireless manipulation of single cells using magnetic microtransporters," in *International Conference on Robotics and Automation*, pp. 2668–2673, 2011.

[34] G.-L. Jiang, Y.-H. Guu, C.-N. Lu, P.-K. Li, H.-M. Shen, L.-S. Lee, J. A. Yeh, and M. T.-K. Hou, "Development of rolling magnetic microrobots," *Journal of Micromechanics and Microengineering*, vol. 20, p. 085042, Aug. 2010.

[35] R. Pelrine, A. Wong-Foy, B. McCoy, D. Holeman, R. Mahoney, G. Myers, J. Herson, and T. Low, "Diamagnetically levitated robots: An approach to massively parallel robotic systems with unusual motion properties," in *IEEE International Conference on Robotics and Automation*, pp. 739–744, 2012.

[36] A. Snezhko and I. S. Aranson, "Magnetic manipulation of self-assembled colloidal asters," *Nature Materials*, vol. 10, pp. 1–6, 2011.

[37] W. Jing, X. Chen, S. Lyttle, and Z. Fu, "A magnetic thin film microrobot with two operating modes," in *International conference on Robotics and Automation*, pp. 96–101, 2011.

[38] B. R. Donald, C. G. Levey, and I. Paprotny, "Planar microassembly by parallel actuation of MEMS microrobots," *Journal of Microelectromechanical Systems*, vol. 17, no. 4, pp. 789–808, 2008.

[39] M. S. Sakar, E. B. Steager, D. H. Kim, a. a. Julius, M. Kim, V. Kumar, and G. J. Pappas, "Modeling, control and experimental characterization of microbiorobots," *International Journal of Robotics Research*, vol. 30, pp. 647–658, Jan. 2011.

[40] N. Chaillet and S. Régnier, "First experiments on MagPieR: A planar wireless magnetic and piezoelectric microrobot," in *IEEE International Conference on Robotics and Automation*, pp. 102–108, 2011.

[41] E. Schaler, M. Tellers, A. Gerratt, I. Penskiy, and S. Bergbreiter, "Toward fluidic microrobots using electrowetting," pp. 3461–3466, 2012.

[42] W. Duan, W. Wang, S. Das, V. Yadav, T. E. Mallouk, and A. Sen, "Synthetic nano-and micromachines in analytical chemistry: Sensing, migration, capture, delivery, and separation," *Annual Review of Analytical Chemistry*, vol. 8, pp. 311–333, 2015.

[43] G. Hwang, R. Braive, L. Couraud, A. Cavanna, O. Abdelkarim, I. Robert-Philip, A. Beveratos, I. Sagnes, S. Haliyo, and S. Regnier, "Electro-osmotic propulsion of helical nanobelt swimmers," *International Journal of Robotics Research*, vol. 30, pp. 806–819, June 2011.

[44] A. Yamazaki, M. Sendoh, K. Ishiyama, K. Ichi Arai, R. Kato, M. Nakano, and H. Fukunaga, "Wireless micro swimming machine with magnetic thin film," *Journal of Magnetism and Magnetic Materials*, vol. 272, pp. E1741–E1742, 2004.

[45] S. Tottori, L. Zhang, F. Qiu, K. K. Krawczyk, A. Franco-Obregón, and B. J. Nelson, "Magnetic helical micromachines: Fabrication, controlled swimming, and cargo transport," *Advanced Materials*, vol. 24, pp. 811–816, Feb. 2012.

[46] L. Zhang, J. J. Abbott, L. Dong, K. E. Peyer, B. E. Kratochvil, H. Zhang, C. Bergeles, and B. J. Nelson, "Characterizing the swimming properties of artificial bacterial flagella," *Nano Letters*, vol. 9, pp. 3663–3667, Oct. 2009.

[47] S. Martel, O. Felfoul, J.-B. Mathieu, A. Chanu, S. Tamaz, M. Mohammadi, M. Mankiewicz, and N. Tabatabaei, "MRI-based medical nanorobotic platform for the control of magnetic nanoparticles and flagellated bacteria for target interventions in human capillaries," *International Journal of Robotics Research*, vol. 28, pp. 1169–1182, Sept. 2009.

[48] D. Hyung Kim, P. Seung Soo Kim, A. Agung Julius, and M. Jun Kim, "Three-dimensional control of Tetrahymena pyriformis using artificial magnetotaxis," *Applied Physics Letters*, vol. 100, no. 5, p. 053702, 2012.

[49] B. J. Williams, S. V. Anand, J. Rajagopalan, and T. A. Saif, "A self-propelled biohybrid swimmer at low Reynolds number," *Nature Communications*, vol. 5, 2014.

[50] V. Arabagi, B. Behkam, E. Cheung, and M. Sitti, "Modeling of stochastic motion of bacteria propelled spherical microbeads," *Journal of Applied Physics*, vol. 109, no. 11, p. 114702, 2011.

[51] T. G. Leong, C. L. Randall, B. R. Benson, N. Bassik, G. M. Stern, and D. H. Gracias, "Tetherless thermobiochemically actuated microgrippers," *Proceedings of the National Academy of Sciences USA*, vol. 106, no. 3, pp. 703–708, 2009.

[52] R. W. Carlsen and M. Sitti, "Bio-hybrid cell-based actuators for microsystems," *Small*, vol. 10, no. 19, pp. 3831–3851, 2014.

[53] M. R. Edwards, R. W. Carlsen, and M. Sitti, "Near and far-wall effects on the three-dimensional motion of bacteria-driven microbeads," *Applied Physics Letters*, vol. 102, p. 143701, 2013.

[54] Y. S. Song and M. Sitti, "Surface-tension-driven biologically inspired water strider robots: Theory and experiments," *IEEE Transactions on Robotics*, vol. 23, no. 3, pp. 578–589, 2007.

[55] M. Sitti, H. Ceylan, W. Hu, J. Giltinan, M. Turan, S. Yim, and E. Diller, "Biomedical applications of untethered mobile milli/microrobots," *Proceedings of the IEEE*, vol. 103, no. 2, pp. 205–224, 2015.

[56] C. Pawashe, S. Floyd, and M. Sitti, "Multiple magnetic microrobot control using electrostatic anchoring," *Applied Physics Letters*, vol. 94, no. 16, p. 164108, 2009.

[57] E. Diller, S. Miyashita, and M. Sitti, "Magnetic hysteresis for multi-state addressable magnetic microrobotic control," in *International Conference on Intelligent Robots and Systems*, pp. 2325–2331, 2012.

[58] C. Pawashe, S. Floyd, E. Diller, and M. Sitti, "Two-dimensional autonomous microparticle manipulation strategies for magnetic microrobots in fluidic environments," *IEEE Transactions on Robotics*, vol. 28, no. 2, pp. 467–477, 2012.

[59] E. Diller, Z. Ye, and M. Sitti, "Rotating magnetic micro-robots for versatile non-contact fluidic manipulation of micro-objects," in *IEEE International Conference on Robots and Systems*, pp. 1291–1296, 2011.

[60] Z. Ye and M. Sitti, "Dynamic trapping and two-dimensional transport of swimming microorganisms using a rotating magnetic microrobot," *Lab on a Chip*, vol. 14, no. 13, pp. 2177–2182, 2014.

[61] E. Diller, C. Pawashe, S. Floyd, and M. Sitti, "Assembly and disassembly of magnetic mobile micro-robots towards deterministic 2-D reconfigurable micro-systems," *International Journal of Robotics Research*, vol. 30, pp. 1667–1680, Sept. 2011.

[62] J. Israelachivili, *Intermolecular and Surface Forces*. Academic Press, 1992.

[63] R. B. Goldfarb and F. R. Fickett, *Units for Magnetic Properties*. US Department of Commerce, National Bureau of Standards, 1985.

[64] B. Cullity and C. Graham, *Introduction to Magnetic Materials*. Wiley-IEEE Press, 2008.

[65] E. Diller and M. Sitti, "Micro-scale mobile robotics," *Foundations and Trends in Robotics*, vol. 2, no. 3, pp. 143–259, 2013.

[66] M. Sitti, "Microscale and nanoscale robotics systems: Characteristics, state of the art, and grand challenges," *IEEE Robotics and Automation Magazine*, vol. 14, no. 1, pp. 53–60, 2007.

[67] M. Sitti, "Voyage of the microrobots," *Nature*, vol. 458, pp. 1121–1122, 2009.

[68] A. Vikram Singh and M. Sitti, "Targeted drug delivery and imaging using mobile milli/microrobots: A promising future towards theranostic pharmaceutical design," *Current Pharmaceutical Design*, vol. 22, no. 11, pp. 1418–1428, 2016.

[69] J. Edd, S. Payen, B. Rubinsky, M. Stoller, and M. Sitti, "Biomimetic propulsion for a swimming surgical micro-robot," in *International Conference on Intelligent Robots and Systems*, pp. 2583–2588, 2003.

[70] B. Behkam and M. Sitti, "Bacterial flagella-based propulsion and on/off motion control of microscale objects," *Applied Physics Letters*, vol. 90, p. 023902, 2007.

[71] K. B. Yesin, K. Vollmers, and B. J. Nelson, "Modeling and control of untethered biomicrorobots in a fluidic environment using electromagnetic fields," *International Journal of Robotics Research*, vol. 25, no. 5-6, pp. 527–536, 2006.

[72] S. Martel, C. Tremblay, S. Ngakeng, and G. Langlois, "Controlled manipulation and actuation of micro-objects with magnetotactic bacteria," *Applied Physics Letters*, vol. 89, p. 233904, 2006.

[73] J. J. Gorman, C. D. McGray, and R. A. Allen, "Mobile microrobot characterization through performance-based competitions," in *Workshop on Performance Metrics for Intelligent Systems*, pp. 122–126, 2009.

[74] B. Behkam and M. Sitti, "Design methodology for biomimetic propulsion of miniature swimming robots," *Journal of Dynamic Systems, Measurement, and Control*, vol. 128, no. 1, p. 36, 2006.

[75] D. L. Hu, B. Chan, and J. W. M. Bush, "The hydrodynamics of water strider locomotion," *Nature*, vol. 424, pp. 663–666, Aug. 2003.

[76] G. P. Sutton and M. Burrows, "Biomechanics of jumping in the flea," *The Journal of Experimental Biology*, vol. 214, pp. 836–847, Mar. 2011.

[77] K. Schmidt-Nielsen, *Scaling: Why Is Animal Size So Important?* Cambridge University Press, 1984.

[78] M. H. Dickinson, F. O. Lehmann, and S. P. Sane, "Wing rotation and the aerodynamic basis of insect flight," *Science*, vol. 284, no. 5422, pp. 1954–1960, 1999.

[79] S. Floyd and M. Sitti, "Design and development of the lifting and propulsion mechanism for a biologically inspired water runner robot," *IEEE Transactions on Robotics*, vol. 24, pp. 698–709, 2008.

[80] M. J. Madou, *Fundamentals of Microfabrication: The Science of Miniaturization*. CRC Press, 2002.

[81] O. Cugat, J. Delamare, and G. Reyne, "Magnetic micro-actuators and systems (MAGMAS)," *IEEE Transactions on Magnetics*, vol. 39, pp. 3607–3612, Nov. 2003.

[82] J. Abbott, Z. Nagy, F. Beyeler, and B. Nelson, "Robotics in the small, part I: Microbotics," *Robotics & Automation Magazine*, vol. 14, pp. 92–103, June 2007.

[83] M. Wautelet, "Scaling laws in the macro-, micro-and nanoworlds," *European Journal of Physics*, vol. 22, pp. 601–611, 2001.

[84] TWI Ltd, http://www.twi-global.com/technical-knowledge/faqs/material-faqs/faq-what-are-the-typical-values-of-surface-energy-for-materials-and-adhesives/, "Table of surface energy values of solid materials."

[85] A. Bahramian and A. Danesh, "Prediction of solid-water-hydrocarbon contact angle," *Journal of Colloid and Interface Science*, vol. 311, pp. 579–586, 2007.

[86] Diversified Enterprises, https://www.accudynetest.com/visc_table.html, "Viscosity, Surface Tension, Specific Density and Molecular Weight of Selected Liquids."

[87] Diversified Enterprises, https://www.accudynetest.com/surface_tension_table.html, "Surface Tension Components and Molecular Weight of Selected Liquids."

[88] Power Chemical Corporation, http://www.powerchemical.net/library/Silicone_Oil.pdf, "Silicone Oil."

[89] M. D. Dickey, R. C. Chiechi, R. J. Larsen, E. A. Weiss, D. A. Weitz, and G. M. Whitesides, "Eutectic gallium-indium (EGaIn): A liquid metal alloy for the formation of stable structures in microchannels at room temperature," *Advanced Functional Materials*, vol. 18, pp. 1097–1104, 2008.

[90] L. Vitos, A. Ruban, H. L. Skriver, and J. Kollar, "The surface energy of metals," *Surface Science*, vol. 411, no. 1, pp. 186–202, 1998.

[91] Z. Yuan, J. Xiao, C. Wang, J. Zeng, S. Xing, and J. Liu, "Preparation of a superamphiphobic surface on a common cast iron substrate," *Journal of Coatings Technology Research*, vol. 8, pp. 773–777, 2011.

[92] P. M. Kulal, D. P. Dubal, C. D. Lokhande, and V. J. Fulari, "Chemical synthesis of Fe2O3 thin films for supercapacitor application," *Journal of Alloys and Compounds*, vol. 509, pp. 2567–2571, 2011.

[93] M. Novotný and K. D. Bartle, "Surface chemistry of glass open tubular (capillary) columns used in gas-liquid chromatography," *Chromatographia*, vol. 7, pp. 122–127, 1974.

[94] A. Zdziennicka, K. Szymczyk, and B. Jaczuk, "Correlation between surface free energy of quartz and its wettability by aqueous solutions of nonionic, anionic and cationic surfactants," *Journal of Colloid and Interface Science*, vol. 340, pp. 243–248, 2009.

[95] F. R. De, Boer, R. Boom, W. C. M. Mattens, A. R. Miedema, and A. K. Niessen, *Cohesion in Metals: Transition Metal Alloys*. Amsterdam: North Holland, 1989.

[96] M. Morita, T. Ohmi, E. Hasegawa, M. Kawakami, and K. Suma, "Control factor of native oxide growth on silicon in air or in ultrapure water," *Applied Physics Letters*, vol. 55, pp. 562–564, 1989.

[97] D. Janssen, R. De Palma, S. Verlaak, P. Heremans, and W. Dehaen, "Static solvent contact angle measurements, surface free energy and wettability determination of various self-assembled monolayers on silicon dioxide," *Thin Solid Films*, vol. 515, pp. 1433–1438, 2006.

[98] K. W. Bewig and W. A. Zisman, "The wetting of gold and platinum by water," *The Journal of Physical Chemistry*, vol. 1097, pp. 4238–4242, 1965.

[99] F. E. Bartell and P. H. Cardwell, "Reproducible contact angles on reproducible metal surfaces. I. Contact angles of water against silver and gold," *Journal of the American Chemical Society*, vol. 64, pp. 494–497, 1942.

[100] D. J. Trevoy and H. Johnson, "The water wettability of metal surfaces," *Journal of Physical Chemistry*, vol. 62, no. 3, pp. 833–837, 1958.

[101] D. A. Winesett, H. Ade, J. Sokolov, M. Rafailovich, and S. Zhu, "Substrate dependence of morphology in thin film polymer blends of polystyrene and poly (methyl methacrylate)," *Polymer International*, vol. 49, pp. 458–462, 2000.

[102] F. Walther, P. Davydovskaya, S. Zürcher, M. Kaiser, H. Herberg, A. M. Gigler, and R. W. Stark, "Stability of the hydrophilic behavior of oxygen plasma activated SU-8," *Journal of Micromechanics and Microengineering*, vol. 17, pp. 524–531, Mar. 2007.

[103] Diversified Enterprises, http://www.accudynetest.com/polytable_01.html, "Critical Surface Tension, Surface Free Energy, Contact Angles with Water, and Hansen Solubility Parameters for Various Polymers."

[104] C. P. Tan and H. G. Craighead, "Surface engineering and patterning using parylene for biological applications," *Materials*, vol. 3, pp. 1803–1832, 2010.

[105] Diversified Enterprises, http://www.accudynetest.com/polytable_03.html?sortby=contact_angle, "Critical surface tension and contact angle with water for various polymers."

[106] J. Visser, "On Hamaker constants: A comparison between Hamaker constants and Lifshitz-van der Waals constants," *Advances in Colloid and Interface Science*, vol. 3, no. 4, pp. 331–363, 1972.

[107] M. Sitti and H. Hashimoto, "Controlled pushing of nanoparticles: Modeling and experiments," *IEEE/ASME Transactions on Mechatronics*, vol. 5, pp. 199–211, June 2000.

[108] N. M. Holbrook and M. A. Zwieniecki, "Transporting water to the tops of trees quick study," *Physics Today*, vol. 61, pp. 76–77, 2008.

[109] A. Adamson, *Physical Chemistry of Surfaces*. Wiley-Interscience, 6th ed., 1997.

[110] S. H. Suhr, Y. S. Song, and M. Sitti, "Biologically inspired miniature water strider robot," in *Robotics: Science and Systems I*, pp. 319–326, 2005.

[111] O. Ozcan, H. Wang, J. D. Taylor, and M. Sitti, "STRIDE II: Water strider inspired miniature robot with circular footpads," *International Journal of Advanced Robotic Systems*, vol. 11, no. 85, 2014.

[112] M. Gauthier, S. Regnier, P. Rougeot, and N. Chaillet, "Forces analysis for micromanipulations in dry and liquid media," *Journal of Micromechatronics*, vol. 3, no. 3-4, pp. 389–413, 2006.

[113] B. Gady, D. Schleef, R. Reifenberger, D. Rimai, and L. DeMejo, "Identification of electrostatic and van der Waals interaction forces between a micrometer-size sphere and a flat substrate," *Physical Review. B, Condensed Matter*, vol. 53, pp. 8065–8070, Mar. 1996.

[114] S. K. Lamoreaux, "Demonstration of the Casimir force in the 0.6 to 6 μm range," *Phys. Rev. Lett.*, vol. 78, no. 1, 1997.

[115] S. Floyd, C. Pawashe, and M. Sitti, "Two-dimensional contact and noncontact micromanipulation in liquid using an untethered mobile magnetic microrobot," *IEEE Transactions on Robotics*, vol. 25, no. 6, pp. 1332–1342, 2009.

[116] M. Sitti and H. Hashimoto, "Teleoperated touch feedback from the surfaces at the nanoscale: modeling and experiments," *IEEE/ASME Transactions on Mechatronics*, vol. 8, no. 2, pp. 287–298, 2003.

[117] M. Sitti and H. Hashimoto, "Tele-nanorobotics using atomic force microscope," in *IEEE/RSJ International Conference on Intelligent Robots and Systems*, pp. 1739–1746, 1998.

[118] M. Sitti and H. Hashimoto, "Tele-nanorobotics using an atomic force microscope as a nanorobot and sensor," *Advanced Robotics*, vol. 13, no. 4, pp. 417–436, 1998.

[119] C. D. Onal and M. Sitti, "Teleoperated 3-D force feedback from the nanoscale with an atomic force microscope," *IEEE Transactions on Nanotechnology*, vol. 9, no. 1, pp. 46–54, 2010.

[120] C. D. Onal, O. Ozcan, and M. Sitti, "Automated 2-D nanoparticle manipulation using atomic force microscopy," *IEEE Transactions on Nanotechnology*, vol. 10, no. 3, pp. 472–481, 2011.

[121] M. Gauthier, S. Regnier, P. Rougeot, N. Chaillet, and S. Régnier, "Analysis of forces for micromanipulations in dry and liquid media," *Journal of Micromechatronics*, vol. 3, pp. 389–413, Sept. 2006.

[122] D. Maugis, "Adhesion of spheres: The JKR-DMT transition using a Dugdale model," *Journal of Colloid and Interface Science*, vol. 150, no. 1, 1992.

[123] O. Piétrement and M. Troyon, "General equations describing elastic indentation depth and normal contact stiffness versus load," *Journal of Colloid and Interface Science*, vol. 226, pp. 166–171, June 2000.

[124] D. Olsen and J. Osteraas, "The critical surface tension of glass," *Journal of Physical Chemistry*, vol. 68, no. 9, pp. 2730–2732, 1964.

[125] S. Rhee, "Surface energies of silicate glasses calculated from their wettability data," *Journal of Materials Science*, vol. 12, pp. 823–824, 1977.

[126] M. Samuelsson and D. Kirchman, "Degradation of adsorbed protein by attached bacteria in relationship to surface hydrophobicity," *American Society for Microbiology*, vol. 56, pp. 3643–3648, 1990.

[127] T. Czerwiec, N. Renevier, and H. Michel, "Low-temperature plasma-assisted nitriding," *Surface and Coatings Technology*, vol. 131, pp. 267–277, 2000.

[128] M. Yuce and A. Demirel, "The effect of nanoparticles on the surface hydrophobicity of polystyrene," *European Physics Journal*, vol. 64, pp. 493–497, 2008.

[129] W. Hu, B. Yang, C. Peng, and S. W. Pang, "Three-dimensional SU-8 structures by reversal UV imprint," *Journal of Vacuum Science & Technology B: Microelectronics and Nanometer Structures*, vol. 24, no. 5, p. 2225, 2006.

[130] P. Hess, "Laser diagnostics of mechanical and elastic properties of silicon and carbon films," *Applied Surface Science*, vol. 106, pp. 429–437, 1996.

[131] M. T. Kim, "Influence of substrates on the elastic reaction of films for the microindentation tests," *Thin Solid Films*, vol. 168, pp. 12–16, 1996.

[132] W. R. Tyson and W. A. Miller, "Surface free energies of solid metals: Estimation from liquid surface tension measurements," *Surface Science*, vol. 62, pp. 267–276, 1977.

[133] H. L. Skriver and N. M. Rosengaard, "Surface energy and work function of elemental metals," *Physical Review B*, vol. 46, no. 11, pp. 7157–7168, 1992.

[134] D. H. Kaelble and J. Moacanin, "A surface energy analysis of bioadhesion," *Polymer*, vol. 18, pp. 475–482, May 1977.

[135] http://www.surface-tension.de/solid-surface-energy.htm, "Solid surface energy data for common polymers."

[136] F. Heslot, A. Cazabat, P. Levinson, and N. Fraysse, "Experiments on wetting on the scale of nanometers: Influence of the surface energy," *Physical Review Letters*, vol. 65, no. 5, pp. 599–602, 1990.

[137] B. Aksak, M. P. Murphy, and M. Sitti, "Adhesion of biologically inspired vertical and angled polymer microfiber arrays," *Langmuir*, vol. 23, no. 6, pp. 3322–3332, 2007.

[138] D. Maugis, *Contact, Adhesion and Rpture of Elastic Solids*. Springer Verlag, 2013.

[139] M. K. Chaudhury, T. Weaver, C. Y. Hui, and E. J. Kramer, "Adhesive contact of cylindrical lens and a flat sheet," *Journal of Applied Physics*, vol. 80, no. 30, 1996.

[140] B. Sumer, C. D. Onal, B. Aksak, and M. Sitti, "An experimental analysis of elliptical adhesive contact," *Journal of Applied Physics*, vol. 107, no. 11, p. 113512, 2010.

[141] K. L. Johnson, "Contact mechanics and adhesion of viscoelastic spheres microstructure and microtribology of polymer surfaces," in *ACS Symposium Series, edited by Tsukruk, V., et al.*, 1999.

[142] U. Abusomwan and M. Sitti, "Effect of retraction speed on adhesion of elastomer fibrillar structures," *Applied Physics Letters*, vol. 101, no. 21, p. 211907, 2012.

[143] M. Sitti, "Atomic force microscope probe based controlled pushing for nanotribological characterization," *IEEE/ASME Transactions on Mechatronics*, vol. 9, no. 2, pp. 343–349, 2004.

[144] M. Sitti, B. Aruk, H. Shintani, and H. Hashimoto, "Scaled teleoperation system for nanoscale interaction and manipulation," *Advanced Robotics*, vol. 17, no. 3, pp. 275–291, 2003.

[145] B. Sümer and M. Sitti, "Rolling and spinning friction characterization of fine particles using lateral force microscopy based contact pushing," *Journal of Adhesion Science and Technology*, vol. 22, pp. 481–506, June 2008.

[146] M. Hagiwara, T. Kawahara, T. Iijima, and F. Arai, "High-speed magnetic microrobot actuation in a microfluidic chip by a fine V-groove surface," *Transactions on Robotics*, vol. 29, no. 2, pp. 363–372, 2013.

[147] C. Dominik and A. G. G. M. Tielens, "An experimental analysis of elliptical adhesive contact," *Astrophys. J.*, vol. 480, pp. 647–673, 1997.

[148] J. A. Williams, "Friction and wear of rotating pivots in MEMS and other small scale devices," *Wear*, vol. 251, pp. 965–972, Oct. 2001.

[149] P. S. Sreetharan, J. P. Whitney, M. D. Strauss, and R. J. Wood, "Monolithic fabrication of millimeter-scale machines," *Journal of Micromechanics and Microengineering*, vol. 22, p. 055027, May 2012.

[150] E. M. Purcell, "Life at low Reynolds number," *American Journal of Physics*, vol. 45, no. 1, pp. 3–11, 1977.

[151] J. Richardson and J. Harker, *Chemical Engineering, Volume 2, 5th Edition*. Butterworth and Heinemann, 2002.

[152] B. Munson, D. Young, and T. Okiishi, *Fundamentals of Fluid Mechanics*. John Wiley and Sons, Inc., 4th ed., 2002.

[153] J. Happel and H. Brenner, *Low Reynolds Number Hydrodynamics*. Prentice-Hall, 1965.

[154] R. Clift, J. R. Grace, and M. E. Weber, *Bubbles, Drops, and Particles*. Academic Press, 2005.

[155] P. A. Valberg and J. P. Butler, "Magnetic particle motions within living cells: Physical theory and techniques," *Biophysical Journal*, vol. 52, pp. 537–550, Oct. 1987.

[156] Q. Liu and A. Prosperetti, "Wall effects on a rotating sphere," *Journal of Fluid Mechanics*, vol. 657, pp. 1–21, 2010.

[157] H. Faxen, "Die Bewegung einer starren Kugel längs der Achse eines mit zäher Flüssigkeit gefüllten Rohres," *Arkiv foer Matematik, Astronomioch Fysik*, vol. 17, no. 27, pp. 1–28, 1923.

[158] A. J. Goldman, R. G. Cox, and H. Brenner, "Slow viscous motion of a sphere parallel to a plane wall-I motion through a quiescent fluid," *Chemical Engineering Science*, vol. 22, no. 4, pp. 637–651, 1967.

[159] P. Bernhard and P. Renaud, "Microstereolithography : Concepts and applications," in *International Conference on Emerging Technologies and Factory Automation*, vol. 00, pp. 289–298, 2001.

[160] P. X. Lan, J. W. Lee, Y.-J. Seol, and D.-W. Cho, "Development of 3D PPF/DEF scaffolds using micro-stereolithography and surface modification," *Journal of Materials Science: Materials in Medicine*, vol. 20, pp. 271–279, Jan. 2009.

[161] J.-W. Choi, E. MacDonald, and R. Wicker, "Multi-material microstereolithography," *International Journal of Advanced Manufacturing Technology*, vol. 49, pp. 543–551, Dec. 2009.

[162] J. Giltinan, E. Diller, C. Mayda, and M. Sitti, "Three-dimensional robotic manipulation and transport of micro-scale objects by a magnetically driven capillary micro-gripper," in *IEEE International Conference on Robotics and Automation*, pp. 2077–2082, 2014.

[163] M. Yasui, M. Ikeuchi, and K. Ikuta, "Magnetic micro actuator with neutral buoyancy and 3D fabrication of cell size magnetized structure," in *IEEE International Conference on Robotics and Automation*, pp. 745–750, 2012.

[164] S. Filiz, L. Xie, L. E. Weiss, and B. Ozdoganlar, "Micromilling of microbarbs for medical implants," *International Journal of Machine Tools and Manufacture*, vol. 48, pp. 459–472, Mar. 2008.

[165] V. Arabagi, L. Hines, and M. Sitti, "Design and manufacturing of a controllable miniature flapping wing robotic platform," *International Journal of Robotics Research*, vol. 31, no. 6, pp. 785–800, 2012.

[166] S. Senturia, *Microsystem Design*. Springer, 2000.

[167] S. Maruo, O. Nakamura, and S. Kawata, "Three-dimensional microfabrication with two-photon-absorbed photopolymerization," *Optics letters*, vol. 22, no. 2, pp. 132–134, 1997.

[168] J. K. Hohmann, M. Renner, E. H. Waller, and G. von Freymann, "Three-dimensional μ-printing: An enabling technology," *Advanced Optical Materials*, vol. 3, no. 11, pp. 1488–1507, 2015.

[169] S. Kim, F. Qiu, S. Kim, A. Ghanbari, C. Moon, L. Zhang, B. J. Nelson, and H. Choi, "Fabrication and characterization of magnetic microrobots for three-dimensional cell culture and targeted transportation," *Advanced Materials*, vol. 25, no. 41, pp. 5863–5868, 2013.

[170] H. Zeng, D. Martella, P. Wasylczyk, G. Cerretti, J.-C. G. Lavocat, C.-H. Ho, C. Parmeggiani, and D. S. Wiersma, "High-resolution 3d direct laser writing for liquid-crystalline elastomer microstructures," *Advanced Materials*, vol. 26, no. 15, pp. 2319–2322, 2014.

[171] H. Ceylan, I. C. Yasa, and M. Sitti, "3D chemical patterning of micromaterials for encoded functionality," *Advanced Materials*, doi: 10.1002/adma.201605072, 2016.

[172] Y. Xia and G. M. Whitesides, "Soft lithography," *Angewandte Chemie*, vol. 37, no. 5, pp. 550–575, 1998.

[173] G. Z. Lum, Z. Ye, X. Dong, H. Marvi, O. Erin, W. Hu, and M. Sitti, "Shape-programmable magnetic soft matter," *Proceedings of the National Academy of Sciences USA*, vol. 113, no. 41, pp. E6007–E6015, 2016.

[174] A. P. Gerratt, I. Penskiy, and S. Bergbreiter, "SOI/elastomer process for energy storage and rapid release," *Journal of Micromechanics and Microengineering*, vol. 20, p. 104011, Oct. 2010.

[175] C. A. Grimes, C. S. Mungle, K. Zeng, M. K. Jain, W. R. Dreschel, M. Paulose, and K. G. Ong, "Wireless magnetoelastic resonance sensors: A criticial review," *Sensors*, vol. 2, pp. 294–313, 2002.

[176] M. K. Jain, Q. Cai, and C. A. Grimes, "A wireless micro-sensor for simultaneous measurement of ph, temperature, and pressure," *Smart Materials and Structures*, vol. 10, pp. 347–353, 2001.

[177] S. Li, L. Fu, J. M. Barbaree, and Z. Y. Cheng, "Resonance behavior of magnetostrictive micro/milli-cantilever and its application as a biosensor," *Sensors and Actuators B: Chemical*, vol. 137, pp. 692–699, 2009.

[178] G. Wu, R. H. Datar, K. M. Hansen, T. Thundat, R. J. Cote, and A. Majumdar, "Bioassay of prostate-specific antigen (PSA) using microcantilevers," *Nature Biotechnology*, vol. 19, pp. 856–860, 2001.

[179] R. Raiteri, M. Grattarola, H.-J. Butt, and P. Skladal, "Micromechanical cantilever-based biosensors," *Sensors and Actuators B*, vol. 79, pp. 115–126, 2001.

[180] A. M. Moulin, S. J. O'Shea, and M. E. Welland, "Microcantilever-based biosensors," *Ultramicroscopy*, vol. 82, pp. 23–31, 2000.

[181] O. Ergeneman and G. Chatzipirpiridis, "In vitro oxygen sensing using intraocular microrobots," *IEEE Transactions on Biomedical Engineering*, vol. 59, pp. 3104–3109, Sept. 2012.

[182] M. Mastrangeli, S. Abbasi, C. Varel, C. Van Hoof, J. P. Celis, and K. F. Böhringer, "Self-assembly from milli- to nanoscales: Mmethods and applications," *Journal of Micromechanics and Microengineering*, vol. 19, p. 083001, July 2009.

[183] D. J. Filipiak, A. Azam, T. G. Leong, and D. H. Gracias, "Hierarchical self-assembly of complex polyhedral microcontainers," *Journal of Micromechanics and Microengineering*, vol. 19, pp. 1–6, July 2009.

[184] J. S. Randhawa, S. S. Gurbani, M. D. Keung, D. P. Demers, M. R. Leahy-Hoppa, and D. H. Gracias, "Three-dimensional surface current loops in terahertz responsive microarrays," *Applied Physics Letters*, vol. 96, no. 19, p. 191108, 2010.

[185] J. S. Song, S. Lee, S. H. Jung, G. C. Cha, and M. S. Mun, "Improved biocompatibility of parylene-C films prepared by chemical vapor deposition and the subsequent plasma treatment," *Journal of Applied Polymer Science*, vol. 112, no. 6, pp. 3677–3685, 2009.

[186] T. Prodromakis, K. Michelakis, T. Zoumpoulidis, R. Dekker, and C. Toumazou, "Biocompatible encapsulation of CMOS based chemical sensors," in *IEEE Sensors*, pp. 791–794, Oct. 2009.

[187] K. M. Sivaraman, "Functional polypyrrole coatings for wirelessly controlled magnetic microrobots," in *Point-of-Care Healthcare Technologies*, pp. 22–25, 2013.

[188] H. Hinghofer-Szalkay and J. E. Greenleaf, "Continuous monitoring of blood volume changes in humans," *Journal of Applied Physiology*, vol. 63, pp. 1003–1007, Sept. 1987.

[189] J. Black, *Handbook of Biomaterial Properties*. Springer, 1998.

[190] R. Fishman, *Cerebrospinal Fluid in Diseases of the Nervous System*. Saunders, 2nd ed., 1980.

[191] S. Palagi and V. Pensabene, "Design and development of a soft magnetically-propelled swimming microrobot," in *International conference on Robotics and Automation*, pp. 5109–5114, 2011.

[192] P. Jena, E. Diller, J. Giltinan, and M. Sitti, "Neutrally buoyant microrobots for enhanced 3D control," in *International Conference on Intelligent Robots and Systems, Workshop on Magnetically Actuated Multiscale Medical Robots*, 2012.

[193] S. Kuiper and B. Hendriks, "Variable-focus liquid lens for miniature cameras," *Applied Physics Letters*, vol. 85, no. 7, pp. 1128–1130, 2004.

[194] K.-H. Jeong, J. Kim, and L. P. Lee, "Biologically inspired artificial compound eyes," *Science*, vol. 312, no. 5773, pp. 557–561, 2006.

[195] M. Amjadi, K.-U. Kyung, I. Park, and M. Sitti, "Stretchable, skin-mountable, and wearable strain sensors and their potential applications: A review," *Advanced Functional Materials*, vol. 26, pp. 1678–1698, 2016.

[196] M. Amjadi, M. Turan, C. P. Clementson, and M. Sitti, "Parallel microcracks-based ultrasensitive and highly stretchable strain sensors," *ACS Applied Materials & Interfaces*, vol. 8, no. 8, pp. 5618–5626, 2016.

[197] J. Bernstein, S. Cho, A. King, A. Kourepenis, P. Maciel, and M. Weinberg, "A micromachined comb-drive tuning fork rate gyroscope," in *IEEE Int. Conf. on Microelectromechanical Systems*, pp. 143–148, 1993.

[198] Y. Sun, S. N. Fry, D. Potasek, D. J. Bell, and B. J. Nelson, "Characterizing fruit fly flight behavior using a microforce sensor with a new comb-drive configuration," *Journal of Microelectromechanical Systems*, vol. 14, no. 1, pp. 4–11, 2005.

[199] G. Lin, R. E. Palmer, K. S. Pister, and K. P. Roos, "Miniature heart cell force transducer system implemented in mems technology," *Biomedical Engineering, IEEE Transactions on*, vol. 48, no. 9, pp. 996–1006, 2001.

[200] Z. Fan, J. Chen, J. Zou, D. Bullen, C. Liu, and F. Delcomyn, "Design and fabrication of artificial lateral line flow sensors," *Journal of Micromechanics and Microengineering*, vol. 12, no. 5, p. 655, 2002.

[201] N. Yazdi, F. Ayazi, and K. Najafi, "Micromachined inertial sensors," *Proceedings of the IEEE*, vol. 86, no. 8, pp. 1640–1659, 1998.

[202] M. Sitti and H. Hashimoto, "Two-dimensional fine particle positioning under an optical microscope using a piezoresistive cantilever as a manipulator," *Journal of Micromechatronics*, vol. 1, no. 1, pp. 25–48, 2000.

[203] Y. Arntz, J. D. Seelig, H. Lang, J. Zhang, P. Hunziker, J. Ramseyer, E. Meyer, M. Hegner, and C. Gerber, "Label-free protein assay based on a nanomechanical cantilever array," *Nanotechnology*, vol. 14, no. 1, p. 86, 2002.

[204] O. Ergeneman, G. Dogangil, M. P. Kummer, J. J. Abbott, M. K. Nazeeruddin, and B. J. Nelson, "A magnetically controlled wireless optical oxygen sensor for intraocular measurements," *IEEE Sensors Journal*, vol. 8, pp. 2022–2024, 2008.

[205] C. Pawashe, *Untethered Mobile Magnetic Micro-Robots*. PhD thesis, Carnegie Mellon University, 2010.

[206] K. G. Ong, C. S. Mungle, and C. A. Grimes, "Control of a magnetoelastic sensor temperature response by magnetic field tuning," *IEEE Transactions on Magnetics*, vol. 39, pp. 3319–3321, 2003.

[207] P. G. Stoyanov and C. A. Grimes, "A remote query magnetostrictive viscosity sensor," *Sensors and Actuators*, vol. 80, pp. 8–14, 2000.

[208] M. Sitti, D. Campolo, J. Yan, and R. S. Fearing, "Development of PZT and PZN-PT based unimorph actuators for micromechanical flapping mechanisms," in *IEEE International Conference on Robotics and Automation*, vol. 4, pp. 3839–3846, 2001.

[209] J. Yan, S. A. Avadhanula, J. Birch, M. H. Dickinson, M. Sitti, T. Su, and R. S. Fearing, "Wing transmission for a micromechanical flying insect," *Journal of Micromechatronics*, vol. 1, no. 3, pp. 221–237, 2001.

[210] M. Sitti, "Piezoelectrically actuated four-bar mechanism with two flexible links for micromechanical flying insect thorax," *IEEE/ASME Transactions on Mechatronics*, vol. 8, no. 1, pp. 26–36, 2003.

[211] V. Arabagi, L. Hines, and M. Sitti, "A simulation and design tool for a passive rotation flapping wing mechanism," *IEEE/ASME Transactions on Mechatronics*, vol. 18, no. 2, pp. 787–798, 2013.

[212] K. J. Son, V. Kartik, J. A. Wickert, and M. Sitti, "An ultrasonic standing-wave-actuated nano-positioning walking robot: Piezoelectric-metal composite beam modeling," *Journal of Vibration and Control*, vol. 12, no. 12, pp. 1293–1309, 2006.

[213] B. Watson, J. Friend, L. Yeo, and M. Sitti, "Piezoelectric ultrasonic resonant micromotor with a volume of less than 1 mm 3 for use in medical microbots," in *IEEE International Conference on Robotics and Automation*, pp. 2225–2230, 2009.

[214] H. Meng and G. Li, "A review of stimuli-responsive shape memory polymer composites," *Polymer*, vol. 54, no. 9, pp. 2199–2221, 2013.

[215] E. W. Jager, E. Smela, and O. Inganäs, "Microfabricating conjugated polymer actuators," *Science*, vol. 290, no. 5496, pp. 1540–1545, 2000.

[216] R. Pelrine, R. Kornbluh, Q. Pei, and J. Joseph, "High-speed electrically actuated elastomers with strain greater than 100%," *Science*, vol. 287, no. 5454, pp. 836–839, 2000.

[217] L. Hines, K. Petersen, and M. Sitti, "Inflated soft actuators with reversible stable deformations," *Advanced Materials*, vol. 28, no. 19, pp. 3690–3696, 2016.

[218] L. Hines, K. Petersen, G. Z. Lum, and M. Sitti, "Soft actuators for small-scale robotics," *Advanced Materials*, doi: 10.1002/adma.201603483, 2016.

[219] W. C. Tang, M. G. Lim, and R. T. Howe, "Electrostatic comb drive levitation and control method," *Journal of Microelectromechanical Systems*, vol. 1, no. 4, pp. 170–178, 1992.

[220] J. de Vicente, D. J. Klingenberg, and R. Hidalgo-Alvarez, "Magnetorheological fluids: A review," *Soft Matter*, vol. 7, no. 8, pp. 3701–3710, 2011.

[221] M. Amjadi and M. Sitti, "High-performance multiresponsive paper actuators," *ACS Nano*, vol. 10, no. 11, pp. 10202–10210, 2016.

[222] W. Wang, W. Duan, S. Ahmed, T. E. Mallouk, and A. Sen, "Small power: Autonomous nano-and micromotors propelled by self-generated gradients," *Nano Today*, vol. 8, no. 5, pp. 531–554, 2013.

[223] S. Fournier-Bidoz, A. C. Arsenault, I. Manners, and G. A. Ozin, "Synthetic self-propelled nanorotors," *Chem. Commun.*, no. 4, pp. 441–443, 2005.

[224] N. S. Zacharia, Z. S. Sadeq, and G. A. Ozin, "Enhanced speed of bimetallic nanorod motors by surface roughening," *Chemical Communications*, no. 39, pp. 5856–5858, 2009.

[225] U. K. Demirok, R. Laocharoensuk, K. M. Manesh, and J. Wang, "Ultrafast catalytic alloy nanomotors," *Angewandte Chemie International Edition*, vol. 47, no. 48, pp. 9349–9351, 2008.

[226] S. Balasubramanian, D. Kagan, K. M. Manesh, P. Calvo-Marzal, G.-U. Flechsig, and J. Wang, "Thermal modulation of nanomotor movement," *Small*, vol. 5, no. 13, pp. 1569–1574, 2009.

[227] J. L. Anderson, "Colloid transport by interfacial forces," *Annual Review of Fluid Mechanics*, vol. 21, no. 1, pp. 61–99, 1989.

[228] J. Ebel, J. L. Anderson, and D. Prieve, "Diffusiophoresis of latex particles in electrolyte gradients," *Langmuir*, vol. 4, no. 2, pp. 396–406, 1988.

[229] S. Sánchez, L. Soler, and J. Katuri, "Chemically powered micro-and nanomotors," *Angewandte Chemie International Edition*, vol. 54, no. 5, pp. 1414–1444, 2015.

[230] G. Volpe, I. Buttinoni, D. Vogt, H.-J. Kümmerer, and C. Bechinger, "Microswimmers in patterned environments," *Soft Matter*, vol. 7, no. 19, pp. 8810–8815, 2011.

[231] I. Buttinoni, G. Volpe, F. Kümmel, G. Volpe, and C. Bechinger, "Active brownian motion tunable by light," *Journal of Physics: Condensed Matter*, vol. 24, no. 28, p. 284129, 2012.

[232] X. Ma, X. Wang, K. Hahn, and S. Sánchez, "Motion control of urea-powered biocompatible hollow microcapsules," *ACS Nano*, vol. 10, no. 3, pp. 3597–3605, 2016.

[233] R. F. Ismagilov, A. Schwartz, N. Bowden, and G. M. Whitesides, "Autonomous movement and self-assembly," *Angewandte Chemie International Edition*, vol. 41, no. 4, pp. 652–654, 2002.

[234] Z. Fattah, G. Loget, V. Lapeyre, P. Garrigue, C. Warakulwit, J. Limtrakul, L. Bouffier, and A. Kuhn, "Straightforward single-step generation of microswimmers by bipolar electrochemistry," *Electrochimica Acta*, vol. 56, no. 28, pp. 10562–10566, 2011.

[235] W. Gao, M. D'Agostino, V. Garcia-Gradilla, J. Orozco, and J. Wang, "Multi-fuel driven janus micromotors," *Small*, vol. 9, no. 3, pp. 467–471, 2013.

[236] F. Mou, C. Chen, H. Ma, Y. Yin, Q. Wu, and J. Guan, "Self-propelled micromotors driven by the magnesium–water reaction and their hemolytic properties," *Angewandte Chemie*, vol. 125, no. 28, pp. 7349–7353, 2013.

[237] W. Gao, A. Pei, and J. Wang, "Water-driven micromotors," *ACS Nano*, vol. 6, no. 9, pp. 8432–8438, 2012.

[238] L. Baraban, R. Streubel, D. Makarov, L. Han, D. Karnaushenko, O. G. Schmidt, and G. Cuniberti, "Fuel-free locomotion of janus motors: Magnetically induced thermophoresis," *ACS Nano*, vol. 7, no. 2, pp. 1360–1367, 2013.

[239] B. Qian, D. Montiel, A. Bregulla, F. Cichos, and H. Yang, "Harnessing thermal fluctuations for purposeful activities: The manipulation of single micro-swimmers by adaptive photon nudging," *Chemical Science*, vol. 4, no. 4, pp. 1420–1429, 2013.

[240] R. Golestanian, "Collective behavior of thermally active colloids," *Physical Review Letters*, vol. 108, no. 3, p. 038303, 2012.

[241] J. W. Bush and D. L. Hu, "Walking on water: biolocomotion at the interface," *Annu. Rev. Fluid Mech.*, vol. 38, pp. 339–369, 2006.

[242] H. Zhang, W. Duan, L. Liu, and A. Sen, "Depolymerization-powered autonomous motors using biocompatible fuel," *Journal of the American Chemical Society*, vol. 135, no. 42, pp. 15734–15737, 2013.

[243] R. Sharma, S. T. Chang, and O. D. Velev, "Gel-based self-propelling particles get programmed to dance," *Langmuir*, vol. 28, no. 26, pp. 10128–10135, 2012.

[244] N. Bassik, B. T. Abebe, and D. H. Gracias, "Solvent driven motion of lithographically fabricated gels," *Langmuir*, vol. 24, no. 21, pp. 12158–12163, 2008.

[245] T. Mitsumata, K. Ikeda, J. P. Gong, and Y. Osada, "Solvent-driven chemical motor," *Applied Physics Letters*, vol. 73, no. 16, pp. 2366–2368, 1998.

[246] C. Luo, H. Li, and X. Liu, "Propulsion of microboats using isopropyl alcohol as a propellant," *Journal of Micromechanics and Microengineering*, vol. 18, no. 6, p. 067002, 2008.

[247] G. Loget and A. Kuhn, "Electric field-induced chemical locomotion of conducting objects," *Nature Communications*, vol. 2, p. 535, 2011.

[248] S. J. Wang, F. D. Ma, H. Zhao, and N. Wu, "Fuel-free locomotion of janus motors: magnetically induced thermophoresis," *ACS Appl. Mater. Interfaces*, vol. 6, pp. 4560–4569, 2014.

[249] M. Bennet, A. McCarthy, D. Fix, M. R. Edwards, F. Repp, P. Vach, J. W. Dunlop, M. Sitti, G. S. Buller, and S. Klumpp, "Influence of magnetic fields on magneto-aerotaxis," *PLoS One*, vol. 9, no. 7, p. e101150, 2014.

[250] M. R. Edwards, R. W. Carlsen, J. Zhuang, and M. Sitti, "Swimming characterization of *Serratia marcescens* for bio-hybrid micro-robotics," *Journal of Micro-Bio Robotics*, vol. 9, no. 3-4, pp. 47–60, 2014.

[251] A. V. Singh and M. Sitti, "Patterned and specific attachment of bacteria on biohybrid bacteria-driven microswimmers," *Advanced Healthcare Materials*, vol. 5, no. 18, pp. 2325–2331, 2016.

[252] B. Behkam and M. Sitti, "Effect of quantity and configuration of attached bacteria on bacterial propulsion of microbeads," *Applied Physics Letters*, vol. 93, p. 223901, Dec. 2008.

[253] J. Zhuang and M. Sitti, "Chemotaxis of bio-hybrid multiple bacteria-driven microswimmers," *Scientific Reports*, vol. 6, article no: 32135, 2016.

[254] H. Danan, A. Herr, and A. Meyer, "New determinations of the saturation magnetization of nickel and iron," *Journal of Applied Physics*, vol. 39, no. 2, p. 669, 1968.

[255] M. Aus, C. Cheung, and B. Szpunar, "Saturation magnetization of porosity-free nanocrystalline cobalt," *Journal of Materials Science Letters*, vol. 7, pp. 1949–1952, 1998.

[256] J. F. Schenck, "Safety of strong, static magnetic fields," *Journal of Magnetic Resonance Imaging*, vol. 12, pp. 2–19, July 2000.

[257] "Criteria for significant risk investigations of magnetic resonance diagnostic devices," tech. rep., U.S. Food and Drug Administration, 2003.

[258] R. Price, "The AAPM/RSNA physics tutorial for residents: MR imaging safety considerations," *Imaging and Therapeutic Technology*, vol. 19, no. 6, pp. 1641–1651, 1999.

[259] D. W. McRobbie, "Occupational exposure in MRI," *The British Journal of Radiology*, vol. 85, pp. 293–312, Apr. 2012.

[260] D. K. Cheng, *Field and Wave Electromagnetics, 2nd Edition*. Addison-Wesley Publishing Company, Inc., 1992.

[261] W. Frix, G. Karady, and B. Venetz, "Comparison of calibration systems for magnetic field measurement equipment," *IEEE Transactions on Power Delivery*, vol. 9, no. 1, pp. 100–108, 1994.

[262] M. E. Rudd, "Optimum spacing of square and circular coil pairs," *Review of Scientific Instruments*, vol. 39, no. 9, p. 1372, 1968.

[263] A. W. Mahoney, J. C. Sarrazin, E. Bamberg, and J. J. Abbott, "Velocity control with gravity compensation for magnetic helical microswimmers," *Advanced Robotics*, vol. 25, pp. 1007–1028, May 2011.

[264] S. Schuerle, S. Erni, M. Flink, B. E. Kratochvil, and B. J. Nelson, "Three-dimensional magnetic manipulation of micro- and nanostructures for applications in life sciences," *IEEE Transactions on Magnetics*, vol. 49, pp. 321–330, Jan. 2013.

[265] R. Bjork, C. Bahl, A. Smith, and N. Pryds, "Comparison of adjustable permanent magnetic field sources," *Journal of Magnetism and Magnetic Materials*, vol. 322, pp. 3664–3671, Nov. 2010.

[266] A. Petruska and J. Abbott, "Optimal permanent-magnet geometries for dipole field approximation," *IEEE Transactions on Magnetics*, vol. 49, no. 2, pp. 811–819, 2013.

[267] A. Mahoney, D. Cowan, and K. Miller, "Control of untethered magnetically actuated tools using a rotating permanent magnet in any position," in *International conference on Robotics and Automation*, pp. 3375–3380, 2012.

[268] T. W. R. Fountain, P. V. Kailat, and J. J. Abbott, "Wireless control of magnetic helical microrobots using a rotating-permanent-magnet manipulator," in *IEEE International Conference on Robotics and Automation*, pp. 576–581, May 2010.

[269] K. Tsuchida, H. M. García-García, W. J. van der Giessen, E. P. McFadden, M. van der Ent, G. Sianos, H. Meulenbrug, A. T. L. Ong, and P. W. Serruys, "Guidewire navigation in coronary artery stenoses using a novel magnetic navigation system: First clinical experience," *Catheterization and Cardiovascular Interventions*, vol. 67, pp. 356–63, Mar. 2006.

[270] J.-B. Mathieu, G. Beaudoin, and S. Martel, "Method of propulsion of a ferromagnetic core in the cardiovascular system through magnetic gradients generated by an mri system," *IEEE Transactions on Biomedical Engineering*, vol. 53, no. 2, pp. 292–299, 2006.

[271] J. Mathieu and S. Martel, "In vivo validation of a propulsion method for untethered medical microrobots using a clinical magnetic resonance imaging system," in *International Conference on Intelligent Robots and Systems*, pp. 502–508, 2007.

[272] K. Belharet, D. Folio, and A. Ferreira, "Endovascular navigation of a ferromagnetic microrobot using MRI-based predictive control," in *IEEE/RSJ International Conference on Intelligent Robots and Systems*, pp. 2804–2809, Oct. 2010.

[273] L. Arcese, M. Fruchard, F. Beyeler, A. Ferreira, and B. Nelson, "Adaptive backstepping and MEMS force sensor for an MRI-guided microrobot in the vasculature," in *IEEE International Conference on Robotics and Automation*, pp. 4121–4126, 2011.

[274] S. T. Chang, V. N. Paunov, D. N. Petsev, and O. D. Velev, "Remotely powered self-propelling particles and micropumps based on miniature diodes," *Nature Materials*, vol. 6, no. 3, pp. 235–240, 2007.

[275] P. Calvo-Marzal, S. Sattayasamitsathit, S. Balasubramanian, J. R. Windmiller, C. Dao, and J. Wang, "Propulsion of nanowire diodes," *Chemical Communications*, vol. 46, no. 10, pp. 1623–1624, 2010.

[276] S. Fusco, M. S. Sakar, S. Kennedy, C. Peters, R. Bottani, F. Starsich, A. Mao, G. A. Sotiriou, S. Pané, S. E. Pratsinis, and D. Mooney, "An integrated microrobotic platform for on-demand, targeted therapeutic interventions," *Advanced Materials*, vol. 26, no. 6, pp. 952–957, 2014.

[277] W. Hu, K. S. Ishii, Q. Fan, and A. T. Ohta, "Hydrogel microrobots actuated by optically generated vapour bubbles," *Lab on a Chip*, vol. 12, pp. 3821–3826, Aug. 2012.

[278] D. Klarman, D. Andelman, and M. Urbakh, "A model of electrowetting, reversed electrowetting and contact angle saturation," *Langmuir*, vol. 27, no. 10, pp. 6031–6041, 2011.

[279] H.-C. Chang and L. Yeo, *Electrokinetically-Driven Microfluidics and Nanofluidics*. Cambridge University Press, 2009.

[280] J. Gong, S.-K. Fan, and C.-J. Kim, "Portable digital microfluidics platform with active but disposable lab-on-chip," in *IEEE International Conference on Microelectromechanical Systems*, vol. 3, pp. 355–358, 2004.

[281] C. Bouyer, P. Chen, S. Güven, T. T. Demirtaş, T. J. Nieland, F. Padilla, and U. Demirci, "A bio-acoustic levitational (BAL) assembly method for engineering of multilayered, 3D brain-like constructs using human embryonic stem cell derived neuro-progenitors," *Advanced Materials*, vol. 28, no. 1, pp. 161–167, 2016.

[282] P. Chen, S. Güven, O. B. Usta, M. L. Yarmush, and U. Demirci, "Biotunable acoustic node assembly of organoids," *Advanced Healthcare Materials*, vol. 4, no. 13, pp. 1937–1943, 2015.

[283] D. Kagan, M. J. Benchimol, J. C. Claussen, E. Chuluun-Erdene, S. Esener, and J. Wang, "Acoustic droplet vaporization and propulsion of perfluorocarbon-loaded microbullets for targeted tissue penetration and deformation," *Angewandte Chemie*, vol. 124, no. 30, pp. 7637–7640, 2012.

[284] B. R. Donald, C. G. Levey, C. D. McGray, D. Rus, and M. Sinclair, "Power delivery and locomotion of untethered microactuators," *Journal of Microelectromechanical Systems*, vol. 12, no. 6, pp. 947–959, 2003.

[285] K. A. Cook-Chennault, N. Thambi, and A. M. Sastry, "Powering MEMS portable devicesa review of non-regenerative and regenerative power supply systems with special emphasis on piezoelectric energy harvesting systems," *Smart Materials and Structures*, vol. 17, no. 4, p. 043001, 2008.

[286] B. Wang, J. Bates, F. Hart, B. Sales, R. Zuhr, and J. Robertson, "Characterization of thin-film rechargeable lithium batteries with lithium cobalt oxide cathodes," *Journal of The Electrochemical Society*, vol. 143, no. 10, pp. 3203–3213, 1996.

[287] A. Shukla, P. Suresh, B. Sheela, and A. Rajendran, "Biological fuel cells and their applications," *Current Science*, vol. 87, no. 4, pp. 455–468, 2004.

[288] A. Kundu, J. Jang, J. Gil, C. Jung, H. Lee, S.-H. Kim, B. Ku, and Y. Oh, "Micro-fuel cellscurrent development and applications," *Journal of Power Sources*, vol. 170, no. 1, pp. 67–78, 2007.

[289] A. Lai, R. Duggirala, and S. Tin, "Radioisotope powered electrostatic microactuators and electronics," in *International Conference on Solid-State Sensors, Actuators and Microsystems*, pp. 269–273, 2007.

[290] H. Li and A. Lal, "Self-reciprocating radioisotope-powered cantilever," *Journal of Applied Physics*, vol. 92, no. 2, pp. 1122–1127, 2002.

[291] A. Kurs, A. Karalis, R. Moffatt, J. D. Joannopoulos, P. Fisher, and M. Soljacic, "Wireless power transfer via strongly coupled magnetic resonances," *Science*, vol. 317, pp. 83–6, July 2007.

[292] D. Schneider, "Electrons unplugged: Wireless power at a distance is still far away," *IEEE Spectrum*, pp. 34–39, 2010.

[293] S. Takeuchi and I. Shimoyama, "Selective drive of electrostatic actuators using remote inductive powering," *Sensors and Actuators A: Physical*, vol. 95, pp. 269–273, Jan. 2002.

[294] N. Shinohara, "Power without wires," *IEEE Microwave Magazine*, pp. 64–73, 2011.

[295] W. Brown, "The history of power transmission by radio waves," *IEEE Transactions on Microwave Theory and Techniques*, vol. 32, no. 9, pp. 1230–1242, 1984.

[296] J. S. Lee, W. Yim, C. Bae, and K. J. Kim, "Wireless actuation and control of ionic polyer-metal composite actuator using a microwave link," *International Journal of Smart and Nano Materials*, vol. 3, no. 4, pp. 244–262, 2012.

[297] T. Shibata, T. Sasaya, and N. Kawahara, "Development of inpipe microrobot using microwave energy transmission," *Electronics and Communications in Japan (Part II: Electronics)*, vol. 84, pp. 1–8, Nov. 2001.

[298] S. Roundy, D. Steingart, L. Frechette, P. Wright, and J. Rabaey, "Power sources for wireless sensor networks," in *Wireless Sensor Networks*, pp. 1–17, Springer, 2004.

[299] S. Hollar, A. Flynn, S. Bergbreiter, S. Member, and K. S. J. Pister, "Robot leg motion in a planarized-SOI, two-layer poly-Si process," *Journal of Microelectromechanical Systems*, vol. 14, no. 4, pp. 725–740, 2005.

[300] K. B. Lee, "Urine-activated paper batteries for biosystems," *Journal of Micromechanics and Microengineering*, vol. 15, no. 9, p. S210, 2005.

[301] S. Roundy, "On the effectiveness of vibration-based energy harvesting," *Journal of Intelligent Material Systems and Structures*, vol. 16, no. 10, pp. 809–823, 2005.

[302] J. Kymissis, C. Kendall, J. Paradiso, and N. Gershenfeld, "Parasitic power harvesting in shoes," in *IEEE International Conference on Wearable Computing*, pp. 132–139, 1998.

[303] S. B. Horowitz, M. Sheplak, L. N. Cattafesta, and T. Nishida, "A MEMS acoustic energy harvester," *Journal of Micromechanics and Microengineering*, vol. 16, no. 9, pp. S174–S181, 2006.

[304] L. Wang and F. G. Yuan, "Vibration energy harvesting by magnetostrictive material," *Smart Materials and Structures*, vol. 17, p. 045009, Aug. 2008.

[305] S. Meninger, J. O. Mur-Miranda, R. Amirtharajah, A. Chandrakasan, and J. Lang, "Vibration-to-electric energy conversion," in *International Symposium on Low Power Electronics and Design*, pp. 48–53, 1999.

[306] C. Williams, C. Shearwood, M. Harradine, P. Mellor, T. Birch, and R. Yates, "Development of an electromagnetic micro-generator," *IEE Proceedings - Circuits, Devices and Systems*, vol. 148, no. 6, p. 337, 2001.

[307] S. P. Beeby, M. J. Tudor, and N. M. White, "Energy harvesting vibration sources for microsystems applications," *Measurement Science and Technology*, vol. 17, pp. R175–R195, Dec. 2006.

[308] J. Stevens, "Optimized thermal design of small delta T thermoelectric generators," in *34th Intersociety Energy Conversion Engineering Conference*, 1999.

[309] A. Holmes and G. Hong, "Axial-flow microturbine with electromagnetic generator: Design, CFD simulation, and prototype demonstration," in *IEEE International Conference on Micro Electro Mechanical Systems*, pp. 568–571, 2004.

[310] D. Carli, D. Brunelli, D. Bertozzi, and L. Benini, "A high-efficiency wind-flow energy harvester using micro turbine," in *IEEE International Symposium on Power Electronics Electrical Drives Automation and Motion*, pp. 778–783, 2010.

[311] H. D. Akaydin, N. Elvin, and Y. Andreopoulos, "Energy harvesting from highly unsteady fluid flows using piezoelectric materials," *Journal of Intelligent Material Systems and Structures*, vol. 21, no. 13, pp. 1263–1278, 2010.

[312] E. Yeatman, "Advances in power sources for wireless sensor nodes," in *International Workshop on Body Sensor Networks*, pp. 20–21, 2004.

[313] R. M. Alexander, *Principles of Animal Locomotion*. Princeton University Press, 2003.

[314] M. A. Woodward and M. Sitti, "Multimo-bat: A biologically inspired integrated jumping–gliding robot," *International Journal of Robotics Research*, vol. 33, no. 12, pp. 1511–1529, 2014.

[315] M. Sitti, A. Menciassi, A. J. Ijspeert, K. H. Low, and S. Kim, "Survey and introduction to the focused section on bio-inspired mechatronics," *IEEE/ASME Transactions on Mechatronics*, vol. 18, no. 2, pp. 409–418, 2013.

[316] Z. Ye, E. Diller, and M. Sitti, "Micro-manipulation using rotational fluid flows induced by remote magnetic micro-manipulators," *Journal of Applied Physics*, vol. 112, p. 064912, Sept. 2012.

[317] H.-W. Tung, K. E. Peyer, D. F. Sargent, and B. J. Nelson, "Noncontact manipulation using a transversely magnetized rolling robot," *Applied Physics Letters*, vol. 103, no. 11, p. 114101, 2013.

[318] H. Tung and D. Frutiger, "Polymer-based wireless resonant magnetic microrobots," in *IEEE International Conference on Robotics and Automation*, pp. 715–720, 2012.

[319] S. Floyd, C. Pawashe, and M. Sitti, "An untethered magnetically actuated micro-robot capable of motion on arbitrary surfaces," in *IEEE International Conference on Robotics and Automation*, pp. 419–424, 2008.

[320] D. Stewart, "Finite-dimensional contact mechanics," *Philosophical Transactions Mathematical, Physical, and Engineering Sciences*, vol. 359, no. 1789, pp. 2467–2482, 2001.

[321] K. Y. Ma, P. Chirarattananon, S. B. Fuller, and R. J. Wood, "Controlled flight of a biologically inspired, insect-scale robot," *Science*, vol. 340, pp. 603–7, May 2013.

[322] G. Taylor, "Analysis of the swimming of microscopic organisms," *Proceedings of the Royal Society A: Mathematical, Physical and Engineering Sciences*, vol. 209, pp. 447–461, Nov. 1951.

[323] E. Lauga and T. R. Powers, "The hydrodynamics of swimming microorganisms," *Reports on Progress in Physics*, vol. 72, p. 096601, Sept. 2009.

[324] C. Elbuken, M. B. Khamesee, and M. Yavuz, "Design and implementation of a micromanipulation system using a magnetically levitated mems robot," *IEEE/ASME Transactions on Mechatronics*, vol. 14, no. 4, pp. 434–445, 2009.

[325] J. J. Abbott, O. Ergeneman, M. P. Kummer, A. M. Hirt, and B. J. Nelson, "Modeling magnetic torque and force for controlled manipulation of soft-magnetic bodies," *IEEE/ASME International Conference on Advanced Intelliget Mechatronics*, vol. 23, pp. 1–6, Sept. 2007.

[326] K. E. Peyer, S. Tottori, F. Qiu, L. Zhang, and B. J. Nelson, "Magnetic helical micromachines," *Chemistry*, pp. 1–12, Nov. 2012.

[327] K. E. Peyer, L. Zhang, and B. J. Nelson, "Bio-inspired magnetic swimming microroobts for biomedical applications," *Nanoscale*, vol. 5, pp. 1259–1272, 2013.

[328] T. Honda, K. Arai, and K. Ishiyama, "Micro swimming mechanisms propelled by external magnetic fields," *IEEE Transactions on Magnetics*, vol. 32, no. 5, pp. 5085–5087, 1996.

[329] B. Behkam and M. Sitti, "Modeling and testing of a biomimetic flagellar propulsion method for microscale biomedical swimming robots," in *IEEE Conference on Advanced Intelligent Mechanotronics*, pp. 24–28, 2005.

[330] J. Lighthill, "Flagellar Hydrodynamics," *SIAM Review*, vol. 18, no. 2, pp. 161–230, 1976.

[331] Z. Ye, S. Régnier, and M. Sitti, "Rotating magnetic miniature swimming robots with multiple flexible flagella," *IEEE Transactions on Robotics*, vol. 30, no. 1, pp. 3–13, 2014.

[332] C. Wiggins and R. Goldstein, "Flexive and propulsive dynamics of elastica at low Reynolds number," *Physical Review Letters*, vol. 80, pp. 3879–3882, Apr. 1998.

[333] M. Lagomarsino, F. Capuani, and C. Lowe, "A simulation study of the dynamics of a driven filament in an Aristotelian fluid," *Journal of Theoretical Biology*, vol. 224, pp. 215–224, Sept. 2003.

[334] I. S. Khalil, A. F. Tabak, A. Klingner, and M. Sitti, "Magnetic propulsion of robotic sperms at low-reynolds number," *Applied Physics Letters*, vol. 109, no. 3, p. 033701, 2016.

[335] M. Roper, R. Dreyfus, J. Baudry, M. Fermigier, J. Bibette, and H. A. Stone, "On the dynamics of magnetically driven elastic filaments," *Journal of Fluid Mechanics*, vol. 554, p. 167, Apr. 2006.

[336] J. J. Abbott, K. E. Peyer, M. C. Lagomarsino, L. Zhang, L. Dong, I. K. Kaliakatsos, and B. J. Nelson, "How should microrobots swim?," *International Journal of Robotics Research*, vol. 28, p. 1434, July 2009.

[337] J. W. Bush and D. L. Hu, "Walking on water: Biolocomotion at the interface," *Annu. Rev. Fluid Mech.*, vol. 38, pp. 339–369, 2006.

[338] Y. S. Song, S. H. Suhr, and M. Sitti, "Modeling of the supporting legs for designing biomimetic water strider robots," in *IEEE International Conference on Robotics and Automation*, pp. 2303–2310, 2006.

[339] S. Floyd and M. Sitti, "Design and development of the lifting and propulsion mechanism for a biologically inspired water runner robot," *IEEE Trans. on Robotics*, vol. 24, no. 3, pp. 698–709, 2008.

[340] M. Karpelson, J. P. Whitney, G.-Y. Wei, and R. J. Wood, "Energetics of flapping-wing robotic insects: Towards autonomous hovering flight," in *IEEE/RSJ International Conference on Intelligent Robots and Systems*, pp. 1630–1637, 2010.

[341] M. H. Dickinson, F.-O. Lehmann, and S. P. Sane, "Wing rotation and the aerodynamic basis of insect flight," *Science*, vol. 284, no. 5422, pp. 1954–1960, 1999.

[342] J. Yan, R. J. Wood, S. Avadhanula, M. Sitti, and R. S. Fearing, "Towards flapping wing control for a micromechanical flying insect," in *IEEE International Conference on Robotics and Automation*, pp. 3901–3908, 2001.

[343] R. J. Wood, "The first takeoff of a biologically inspired at-scale robotic insect," *IEEE Transactions on Robotics*, vol. 24, no. 2, pp. 341–347, 2008.

[344] K. Y. Ma, P. Chirarattananon, S. B. Fuller, and R. J. Wood, "Controlled flight of a biologically inspired, insect-scale robot," *Science*, vol. 340, no. 6132, pp. 603–607, 2013.

[345] L. Hines, V. Arabagi, and M. Sitti, "Shape memory polymer-based flexure stiffness control in a miniature flapping-wing robot," *IEEE Transactions on Robotics*, vol. 28, no. 4, pp. 987–990, 2012.

[346] L. Hines, D. Colmenares, and M. Sitti, "Platform design and tethered flight of a motor-driven flapping-wing system," in *IEEE International Conference on Robotics and Automation*, pp. 5838–5845, 2015.

[347] "Development of the nano hummingbird: A tailless flapping wing micro air vehicle," in *AIAA Aerospace Sciences Meeting*, pp. 1–24, 2012.

[348] L. Hines, D. Campolo, and M. Sitti, "Liftoff of a motor-driven, flapping-wing microaerial vehicle capable of resonance," *IEEE Transactions on Robotics*, vol. 30, no. 1, pp. 220–232, 2014.

[349] K. Meng, W. Zhang, W. Chen, H. Li, P. Chi, C. Zou, X. Wu, F. Cui, W. Liu, and J. Chen, "The design and micromachining of an electromagnetic MEMS flapping-wing micro air vehicle," *Microsystem Technologies*, vol. 18, no. 1, pp. 127–136, 2012.

[350] J. Bronson, J. Pulskamp, R. Polcawich, C. Kroninger, and E. Wetzel, "PZT MEMS actuated flapping wings for insect-inspired robotics," in *IEEE International Conference on icro Electro Mechanical Systems*, pp. 1047–1050, 2009.

[351] I. Kroo and P. Kunz, "Development of the mesicopter: A miniature autonomous rotorcraft," in *American Helicopter Society (AHS) Vertical Lift Aircraft Design Conference, San Francisco, CA*, 2000.

[352] A. Kushleyev, D. Mellinger, C. Powers, and V. Kumar, "Towards a swarm of agile micro quadrotors," *Autonomous Robots*, vol. 35, no. 4, pp. 287–300, 2013.

[353] K. Fregene, D. Sharp, C. Bolden, J. King, C. Stoneking, and S. Jameson, "Autonomous guidance and control of a biomimetic single-wing MAV," in *AUVSI Unmanned Systems Conference*, pp. 1–12, 2011.

[354] J. Glasheen and T. McMahon, "A hydrodynamic model of locomotion in the basilisk lizard," *Nature*, vol. 380, no. 6572, pp. 340–341, 1996.

[355] S. T. Hsieh and G. V. Lauder, "Running on water: Three-dimensional force generation by basilisk lizards," *Proceedings of the National Academy of Sciences USA*, vol. 101, no. 48, pp. 16784–16788, 2004.

[356] S. Floyd, T. Keegan, J. Palmisano, and M. Sitti, "A novel water running robot inspired by basilisk lizards," in *IEEE/RSJ International Conference on Intelligent Robots and Systems*, pp. 5430–5436, 2006.

[357] H. S. Park, S. Floyd, and M. Sitti, "Roll and pitch motion analysis of a biologically inspired quadruped water runner robot," *International Journal of Robotics Research*, vol. 29, no. 10, pp. 1281–1297, 2010.

[358] C. Bergeles, G. Fagogenis, J. Abbott, and B. Nelson, "Tracking intraocular microdevices based on colorspace evaluation and statistical color/shape information," in *International Conference on Robotics and Automation*, pp. 3934–3939, 2009.

[359] D. Frantz and A. Wiles, "Accuracy assessment protocols for electromagnetic tracking systems," *Physics in Medicine and Biology*, vol. 48, no. 14, pp. 2241–2251, 2003.

[360] V. Schlageter, "Tracking system with five degrees of freedom using a 2D-array of Hall sensors and a permanent magnet," *Sensors and Actuators A: Physical*, vol. 92, pp. 37–42, Aug. 2001.

[361] B. Gimi, D. Artemov, T. Leong, D. H. Gracias, and Z. M. Bhujwalla, "MRI of regular-shaped cell-encapsulating polyhedral microcontainers," *Magnetic Resonance in Medicine*, vol. 58, pp. 1283–7, Dec. 2007.

[362] C. Dahmen, D. Folio, T. Wortmann, A. Kluge, A. Ferreira, and S. Fatikow, "Evaluation of a MRI based propulsion/control system aiming at targeted micro/nano-capsule therapeutics," in *International Conference on Intelligent Robots and Systems*, pp. 2565–2570, 2012.

[363] N. Olamaei, F. Cheriet, G. Beaudoin, and S. Martel, "MRI visualization of a single 15 m navigable imaging agent and future microrobot," in *Annual International Conference of the IEEE Engineering in Medicine and Biology Society*, vol. 2010, pp. 4355–8, Jan. 2010.

[364] S. Webb, ed., *The Physics of Medical Imaging*. CRC Press, 2 ed., 2010.

[365] G. Neumann, P. DePablo, A. Finckh, L. B. Chibnik, F. Wolfe, and J. Duryea, "Patient repositioning reproducibility of joint space width measurements on hand radiographs," *Arthritis Care & Research*, vol. 63, pp. 203–207, Feb. 2011.

[366] J. H. Daniel, A. Sawant, M. Teepe, C. Shih, R. A. Street, and L. E. Antonuk, "Fabrication of high aspect-ratio polymer microstructures for large-area electronic portal X-ray imagers," *Sensors and Actuators. A, Physical*, vol. 140, pp. 185–193, Nov. 2007.

[367] T. Deffieux, J.-L. Gennisson, M. Tanter, and M. Fink, "Assessment of the mechanical properties of the musculoskeletal system using 2-D and 3-D very high frame rate ultrasound," *IEEE Transactions on Ultrasonics, Ferroelectrics, and Frequency Control*, vol. 55, pp. 2177–90, Oct. 2008.

[368] P. N. Wells, "Current status and future technical advances of ultrasonic imaging," *Engineering in Medicine and Biology*, vol. 19, no. 5, pp. 14–20, 2000.

[369] Z. Nagy, M. Fluckiger, and O. Ergeneman, "A wireless acoustic emitter for passive localization in liquids," in *IEEE International Conference on Robotics and Automation*, (Kobe), pp. 2593–2598, 2009.

[370] S. Earnshaw, "On the nature of the molecular forces which regulate the constitution of the luminiferous ether," *Transactions of the Cambridge Philosophical Society*, vol. 7, no. July, pp. 97–112, 1842.

[371] D. H. Kim, E. B. Steager, U. K. Cheang, D. Byun, and M. J. Kim, "A comparison of vision-based tracking schemes for control of microbiorobots," *Journal of Micromechanics and Microengineering*, vol. 20, p. 065006, June 2010.

[372] D. Lowe, "Object recognition from local scale-invariant features," in *IEEE International Conference on Computer Vision*, pp. 1150–1157 vol.2, 1999.

[373] D. G. Lowe, "Distinctive image features from scale-invariant keypoints," *International Journal of Computer Vision*, vol. 60, pp. 91–110, Nov. 2004.

[374] M. Wu, J. W. Roberts, and M. Buckley, "Three-dimensional fluorescent particle tracking at micron-scale using a single camera," *Experiments in Fluids*, vol. 38, pp. 461–465, Feb. 2005.

[375] K. B. Yesin, K. Vollmers, and B. J. Nelson, "Guidance of magnetic intraocular microrobots by active defocused tracking," in *International Conference on Intelligent Robots*, pp. 3309–3314, 2004.

[376] C. Bergeles, B. E. Kratochvil, and B. J. Nelson, "Visually servoing magnetic intraocular microdevices," *IEEE Transactions on Robotics*, vol. 28, no. 4, pp. 798–809, 2012.

[377] K. F. Bohringer, B. R. Donald, and N. C. MacDonald, "Programmable Force Fields for Distributed Manipulation, with Applications to MEMS Actuator Arrays and Vibratory Parts Feeders," *International Journal of Robotics Research*, vol. 18, pp. 168–200, Feb. 1999.

[378] I. Paprotny, C. G. Levey, P. K. Wright, and B. R. Donald, "Turning-rate selective control: A new method for independent control of stress-engineered MEMS microrobots," in *Robotics: Science and Systems*, 2012.

[379] S. Floyd, E. Diller, C. Pawashe, and M. Sitti, "Control methodologies for a heterogeneous group of untethered magnetic micro-robots," *International Journal of Robotics Research*, vol. 30, pp. 1553–1565, Mar. 2011.

[380] E. Diller, S. Floyd, C. Pawashe, and M. Sitti, "Control of multiple heterogeneous magnetic microrobots in two dimensions on nonspecialized surfaces," *IEEE Transactions on Robotics*, vol. 28, no. 1, pp. 172–182, 2012.

[381] P. Vartholomeos, M. R. Akhavan-sharif, and P. E. Dupont, "Motion planning for multiple millimeter-scale magnetic capsules in a fluid environment," in *IEEE Int. Conf. Robotics and Automation*, pp. 1927–1932, 2012.

[382] E. Diller, J. Giltinan, and M. Sitti, "Independent control of multiple magnetic microrobots in three dimensions," *International Journal of Robotics Research*, vol. 32, pp. 614–631, May 2013.

[383] S. Tottori, N. Sugita, R. Kometani, S. Ishihara, and M. Mitsuishi, "Selective control method for multiple magnetic helical microrobots," *Journal of Micro-Nano Mechatronics*, vol. 6, pp. 89–95, Aug. 2011.

[384] E. Diller and S. Miyashita, "Remotely addressable magnetic composite micropumps," *RSC Advances*, vol. 2, pp. 3850–3856, Feb. 2012.

[385] S. Miyashita, E. Diller, and M. Sitti, "Two-dimensional magnetic micro-module reconfigurations based on inter-modular interactions," *International Journal of Robotics Research*, vol. 32, pp. 591–613, May 2013.

[386] P. M. Braillon, "Magnetic plate comprising permanent magnets and electropermanent magnets," 1978. US Patent 4,075,589.

[387] K. Gilpin, A. Knaian, and D. Rus, "Robot pebbles: One centimeter modules for programmable matter through self-disassembly," in *IEEE International Conference on Robotics and Automation*, pp. 2485–2492, 2010.

[388] A. Menciassi, A. Eisinberg, I. Izzo, and P. Dario, "From macro to micro manipulation: Models and experiments," *IEEE/ASME Transactions on Mechatronics*, vol. 9, pp. 311–320, June 2004.

[389] M. Sitti, "Survey of nanomanipulation systems," in *IEEE Conference on Nanotechnology*, pp. 75–80, 2001.

[390] S. Kim, J. Wu, A. Carlson, S. H. Jin, A. Kovalsky, P. Glass, Z. Liu, N. Ahmed, S. L. Elgan, W. Chen, M. F. Placid, M. Sitti, Y. Huang, and J. A. Rogers, "Microstructured elastomeric surfaces with reversible adhesion and examples of their use in deterministic assembly by transfer printing," *Proceedings of the National Academy of Sciences USA*, vol. 107, no. 40, pp. 17095–17100, 2010.

[391] J. Giltinan, E. Diller, and M. Sitti, "Programmable assembly of heterogeneous microparts by an untethered mobile capillary microgripper," *Lab on a Chip*, vol. 16, no. 22, pp. 4445–4457, 2016.

[392] T. Petit, L. Zhang, K. E. Peyer, B. E. Kratochvil, and B. J. Nelson, "Selective trapping and manipulation of microscale objects using mobile microvortices," *Nano Letters*, vol. 12, pp. 156–60, Jan. 2012.

[393] L. Zhang, T. Petit, K. E. Peyer, and B. J. Nelson, "Targeted cargo delivery using a rotating nickel nanowire," *Nanomedicine*, vol. 8, pp. 1074–1080, Mar. 2012.

[394] L. Zhang, K. E. Peyer, and B. J. Nelson, "Artificial bacterial flagella for micromanipulation," *Lab on a Chip*, vol. 10, pp. 2203–15, Sept. 2010.

[395] S. E. Chung, X. Dong, and M. Sitti, "Three-dimensional heterogeneous assembly of coded microgels using an untethered mobile microgripper," *Lab on a Chip*, vol. 15, no. 7, pp. 1667–1676, 2015.

[396] S. Floyd, C. Pawashe, and M. Sitti, "Microparticle manipulation using multiple untethered magnetic micro-robots on an electrostatic surface," in *IEEE/RSJ International Conference on Intelligent Robots and Systems*, pp. 528–533, Oct. 2009.

[397] B. J. Nelson, I. K. Kaliakatsos, and J. J. Abbott, "Microrobots for minimally invasive medicine," *Annual Review of Biomedical Engineering*, vol. 12, pp. 55–85, Aug. 2010.

[398] S. Yim, E. Gultepe, D. H. Gracias, and M. Sitti, "Biopsy using a magnetic capsule endoscope carrying, releasing, and retrieving untethered microgrippers," *IEEE Transactions on Biomedical Engineering*, vol. 61, no. 2, pp. 513–521, 2014.

[399] S. Yim and M. Sitti, "Design and rolling locomotion of a magnetically actuated soft capsule endoscope," *IEEE Transactions on Robotics*, vol. 28, no. 1, pp. 183–194, 2012.

[400] S. Yim and M. Sitti, "Shape-programmable soft capsule robots for semi-implantable drug delivery," *IEEE Transactions on Robotics*, vol. 28, no. 5, pp. 1198–1202, 2012.

[401] S. Yim, K. Goyal, and M. Sitti, "Magnetically actuated soft capsule with the multimodal drug release function," *IEEE/ASME Transactions on Mechatronics*, vol. 18, no. 4, pp. 1413–1418, 2013.

[402] Z. Hosseinidoust, B. Mostaghaci, O. Yasa, B.-W. Park, A. V. Singh, and M. Sitti, "Bioengineered and biohybrid bacteria-based systems for drug delivery," *Advanced Drug Delivery Reviews*, vol. 106, pp. 27–44, 2016.

[403] G. Dogangil and O. Ergeneman, "Toward targeted retinal drug delivery with wireless magnetic microrobots," in *International Conference on Intelligent Robots and Systems*, pp. 1921–1926, 2008.

[404] W. Gao and J. Wang, "The environmental impact of micro/nanomachines: A review," *ACS Nano*, vol. 8, no. 4, pp. 3170–3180, 2014.

[405] J. Parmar, D. Vilela, E. Pellicer, D. Esqué-de los Ojos, J. Sort, and S. Sánchez, "Reusable and long-lasting active microcleaners for heterogeneous water remediation," *Advanced Functional Materials*, 2016.

[406] M. Yim, W.-M. Shen, B. Salemi, D. Rus, M. Moll, H. Lipson, E. Klavins, and G. S. Chirikjian, "Modular Self-Reconfigurable Robot Systems," *IEEE Robotics and Automation Magazine*, vol. 14, no. 1, pp. 43–52, 2007.

[407] W.-M. Shen, H. Chiu, M. Rubenstein, and B. Salemi, "Rolling and Climbing by the Multifunctional SuperBot Reconfigurable Robotic System," in *Proceedings of the Space Technology International Forum*, (Albuquerque, New Mexico), pp. 839–848, 2008.

[408] S. C. Goldstein, J. D. Campbell, and T. C. Mowry, "Programmable matter," *Computer*, vol. 38, no. 6, pp. 99–101, 2005.

[409] E. Yoshida, S. Kokaji, S. Murata, K. Tomita, and H. Kurokawa, "Micro Self-reconfigurable Robot using Shape Memory Alloy," *Journal of Robotics and Mechatronics*, vol. 13, no. 2, pp. 212–219, 2001.

[410] M. T. Tolley, M. Krishnan, D. Erickson, and H. Lipson, "Dynamically programmable fluidic assembly," *Applied Physics Letters*, vol. 93, no. 25, p. 254105, 2008.

[411] M. Kalontarov, M. T. Tolley, H. Lipson, and D. Erickson, "Hydrodynamically driven docking of blocks for 3D fluidic assembly," *Microfluidics and Nanofluidics*, vol. 9, pp. 551–558, Feb. 2010.

[412] K. Autumn, M. Sitti, Y. A. Liang, A. M. Peattie, W. R. Hansen, S. Sponberg, T. W. Kenny, R. Fearing, J. N. Israelachvili, and R. J. Full, "Evidence for van der waals adhesion in gecko setae," *Proceedings of the National Academy of Sciences USA*, vol. 99, no. 19, pp. 12252–12256, 2002.

[413] M. Sitti and R. S. Fearing, "Synthetic gecko foot-hair micro/nano-structures as dry adhesives," *Journal of Adhesion Science and Technology*, vol. 17, no. 8, pp. 1055–1073, 2003.

[414] M. P. Murphy, B. Aksak, and M. Sitti, "Gecko-inspired directional and controllable adhesion," *Small*, vol. 5, no. 2, pp. 170–175, 2009.

[415] M. P. Murphy, S. Kim, and M. Sitti, "Enhanced adhesion by gecko-inspired hierarchical fibrillar adhesives," *ACS Applied Materials & Interfaces*, vol. 1, no. 4, pp. 849–855, 2009.

[416] S. Kim and M. Sitti, "Biologically inspired polymer microfibers with spatulate tips as repeatable fibrillar adhesives," *Applied Physics Letters*, vol. 89, no. 26, p. 261911, 2006.

[417] G. J. Shah and M. Sitti, "Modeling and design of biomimetic adhesives inspired by gecko foot-hairs," in *IEEE International Conference on Robotics and Biomimetics*, pp. 873–878, 2004.

[418] Y. Mengüç, S. Y. Yang, S. Kim, J. A. Rogers, and M. Sitti, "Gecko-inspired controllable adhesive structures applied to micromanipulation," *Advanced Functional Materials*, vol. 22, no. 6, pp. 1246–1254, 2012.

[419] B. Aksak, M. Sitti, A. Cassell, J. Li, M. Meyyappan, and P. Callen, "Friction of partially embedded vertically aligned carbon nanofibers inside elastomers," *Applied Physics Letters*, vol. 91, no. 6, pp. 61906–61906, 2007.

[420] S. Kim, M. Sitti, C.-Y. Hui, R. Long, and A. Jagota, "Effect of backing layer thickness on adhesion of single-level elastomer fiber arrays," *Applied Physics Letters*, vol. 91, no. 16, pp. 161905–161905, 2007.

[421] S. Kim, E. Cheung, and M. Sitti, "Wet self-cleaning of biologically inspired elastomer mushroom shaped microfibrillar adhesives," *Langmuir*, vol. 25, no. 13, pp. 7196–7199, 2009.

[422] M. P. Murphy and M. Sitti, "Waalbot: An agile small-scale wall-climbing robot utilizing dry elastomer adhesives," *IEEE/ASME Transactions on Mechatronics*, vol. 12, no. 3, pp. 330–338, 2007.

[423] P. Glass, H. Chung, N. R. Washburn, and M. Sitti, "Enhanced reversible adhesion of dopamine methacrylamide-coated elastomer microfibrillar structures under wet conditions," *Langmuir*, vol. 25, no. 12, pp. 6607–6612, 2009.

[424] M. P. Murphy, B. Aksak, and M. Sitti, "Adhesion and anisotropic friction enhancements of angled heterogeneous micro-fiber arrays with spherical and spatula tips," *Journal of Adhesion Science and Technology*, vol. 21, no. 12-13, pp. 1281–1296, 2007.

[425] S. Kim, M. Sitti, T. Xie, and X. Xiao, "Reversible dry micro-fibrillar adhesives with thermally controllable adhesion," *Soft Matter*, vol. 5, no. 19, pp. 3689–3693, 2009.

[426] P. Glass, H. Chung, N. R. Washburn, and M. Sitti, "Enhanced wet adhesion and shear of elastomeric micro-fiber arrays with mushroom tip geometry and a photopolymerized p (dma-co-mea) tip coating," *Langmuir*, vol. 26, no. 22, pp. 17357–17362, 2010.

[427] S. Kim, B. Aksak, and M. Sitti, "Enhanced friction of elastomer microfiber adhesives with spatulate tips," *Applied Physics Letters*, vol. 91, no. 22, p. 221913, 2007.

[428] E. Cheung and M. Sitti, "Adhesion of biologically inspired polymer microfibers on soft surfaces," *Langmuir*, vol. 25, no. 12, pp. 6613–6616, 2009.

[429] M. Sitti and R. S. Fearing, "Synthetic gecko foot-hair micro/nano-structures for future wall-climbing robots," in *IEEE International Conference on Robotics and Automation*, vol. 1, pp. 1164–1170, 2003.

[430] Y. Mengüç, M. Röhrig, U. Abusomwan, H. Hölscher, and M. Sitti, "Staying sticky: contact self-cleaning of gecko-inspired adhesives," *Journal of The Royal Society Interface*, vol. 11, no. 94, p. 20131205, 2014.

[431] M. Sitti, B. Cusick, B. Aksak, A. Nese, H.-i. Lee, H. Dong, T. Kowalewski, and K. Matyjaszewski, "Dangling chain elastomers as repeatable fibrillar adhesives," *ACS Applied Materials & Interfaces*, vol. 1, no. 10, pp. 2277–2287, 2009.

[432] R. Long, C.-Y. Hui, S. Kim, and M. Sitti, "Modeling the soft backing layer thickness effect on adhesion of elastic microfiber arrays," *Journal of Applied Physics*, vol. 104, no. 4, p. 044301, 2008.

[433] B. Aksak, C.-Y. Hui, and M. Sitti, "The effect of aspect ratio on adhesion and stiffness for soft elastic fibres," *Journal of The Royal Society Interface*, vol. 8, no. 61, pp. 1166–1175, 2011.

[434] M. Piccardo, A. Chateauminois, C. Fretigny, N. M. Pugno, and M. Sitti, "Contact compliance effects in the frictional response of bioinspired fibrillar adhesives," *Journal of The Royal Society Interface*, vol. 10, no. 83, p. 20130182, 2013.

[435] S. Song and M. Sitti, "Soft grippers using micro-fibrillar adhesives for transfer printing," *Advanced Materials*, vol. 26, no. 28, pp. 4901–4906, 2014.

[436] B. Aksak, K. Sahin, and M. Sitti, "The optimal shape of elastomer mushroom-like fibers for high and robust adhesion," *Beilstein Journal of Nanotechnology*, vol. 5, no. 1, pp. 630–638, 2014.

[437] E. Diller, N. Zhang, and M. Sitti, "Modular micro-robotic assembly through magnetic actuation and thermal bonding," *Journal of Micro-Bio Robotics*, vol. 8, no. 3-4, pp. 121–131, 2013.

[438] Z. Ye, G. Z. Lum, S. Song, S. Rich, and M. Sitti, "Phase change of gallium enables highly reversible and switchable adhesion," *Advanced Materials*, vol. 28, no. 25, pp. 5088–5092, 2016.

[439] A.-L. Routier-Kierzkowska, A. Weber, P. Kochova, D. Felekis, B. J. Nelson, C. Kuhlemeier, and R. S. Smith, "Cellular force microscopy for in vivo measurements of plant tissue mechanics," *Plant Physiology*, vol. 158, pp. 1514–22, Apr. 2012.

[440] D. Felekis, S. Muntwyler, H. Vogler, F. Beyeler, U. Grossniklaus, and B. Nelson, "Quantifying growth mechanics of living, growing plant cells in situ using microrobotics," *Micro & Nano Letters*, vol. 6, no. 5, p. 311, 2011.

[441] T. Kawahara and M. Sugita, "On-chip manipulation and sensing of microorganisms by magnetically driven microtools with a force sensing structure," in *International conference on Robotics and Automation*, pp. 4112–4117, 2012.

[442] T. Braun, V. Barwich, M. Ghatkesar, A. Bredekamp, C. Gerber, M. Hegner, and H. Lang, "Micromechanical mass sensors for biomolecular detection in a physiological environment," *Physical Review E*, vol. 72, no. 3, pp. 1–9, 2005.

[443] M. Jain and C. Grimes, "A wireless magnetoelastic micro-sensor array for simultaneous measurement of temperature and pressure," *IEEE Transactions on Magnetics*, vol. 37, pp. 2022–2024, July 2001.

Index

Intelligent Robotics and Autonomous Agents

Edited by Ronald C. Arkin

Dorigo, Marco, and Marco Colombetti, *Robot Shaping: An Experiment in Behavior Engineering*

Arkin, Ronald C., *Behavior-Based Robotics*

Stone, Peter, *Layered Learning in Multiagent Systems: A Winning Approach to Robotic Soccer*

Wooldridge, Michael, *Reasoning About Rational Agents*

Murphy, Robin R., *An Introduction to AI Robotics*

Mason, Matthew T., *Mechanics of Robotic Manipulation*

Kraus, Sarit, *Strategic Negotiation in Multiagent Environments*

Nolfi, Stefano, and Dario Floreano, *Evolutionary Robotics: The Biology, Intelligence, and Technology of Self-Organizing Machines*

Siegwart, Roland, and Illah R. Nourbakhsh, *Introduction to Autonomous Mobile Robots*

Breazeal, Cynthia L., *Designing Sociable Robots*

Bekey, George A., *Autonomous Robots: From Biological Inspiration to Implementation and Control*

Choset, Howie, Kevin M. Lynch, Seth Hutchinson, George Kantor, Wolfram Burgard, Lydia E. Kavraki, and Sebastian Thrun, *Principles of Robot Motion: Theory, Algorithms, and Implementations*

Thrun, Sebastian, Wolfram Burgard, and Dieter Fox, *Probabilistic Robotics*

Mataric, Maja J., *The Robotics Primer*

Wellman, Michael P., Amy Greenwald, and Peter Stone, *Autonomous Bidding Agents: Strategies and Lessons from the Trading Agent Competition*

Floreano, Dario and Claudio Mattiussi, *Bio-Inspired Artificial Intelligence: Theories, Methods, and Technologies*

Sterling, Leon S. and Kuldar Taveter, *The Art of Agent-Oriented Modeling*

Stoy, Kasper, David Brandt, and David J. Christensen, *An Introduction to Self-Reconfigurable Robots*

Lin, Patrick, Keith Abney, and George A. Bekey, editors, *Robot Ethics: The Ethical and Social Implications of Robotics*

Weiss, Gerhard, editor, *Multiagent Systems, second edition*

Vargas, Patricia A., Ezequiel A. Di Paolo, Inman Harvey, and Phil Husbands, editors, *The Horizons of Evolutionary Robotics*

Murphy, Robin R., *Disaster Robotics*

Cangelosi, Angelo and Matthew Schlesinger, *Developmental Robotics: From Babies to Robots*

Everett, H. R., *Unmanned Systems of World Wars I and II*

Sitti, Metin, *Mobile Microrobotics*